Making Sense in Common

CARY WOLFE, SERIES EDITOR

66 *Making Sense in Common: A Reading of Whitehead in Times of Collapse*
    Isabelle Stengers

65 *Our Grateful Dead: Stories of Those Left Behind*
    Vinciane Despret

64 *Prosthesis*
    David Wills

63 *Molecular Capture: The Animation of Biology*
    Adam Nocek

62 *Clang*
    Jacques Derrida

61 *Radioactive Ghosts*
    Gabriele Schwab

60 *Gaian Systems: Lynn Margulis, Neocybernetics, and the End of the Anthropocene*
    Bruce Clarke

59 *The Probiotic Planet: Using Life to Manage Life*
    Jamie Lorimer

58 *Individuation in Light of Notions of Form and Information. Volume II: Supplemental Texts*
    Gilbert Simondon

57 *Individuation in Light of Notions of Form and Information*
    Gilbert Simondon

56 *Thinking Plant Animal Human: Encounters with Communities of Difference*
    David Wood

55 *The Elements of Foucault*
    Gregg Lambert

(continued on page 200)

Making Sense in Common

A Reading of Whitehead in Times of Collapse

Isabelle Stengers

Translated and with an Introduction by Thomas Lamarre

posthumanities **66**

University of Minnesota Press
Minneapolis
London

The University of Minnesota Press gratefully acknowledges the generous assistance provided for the publication of this book by the Hamilton P. Traub University Press Fund.

Originally published in French as *Reactiver le sens commun: Lecture de Whitehead en temps de débâcle.* Copyright Editions La Découverte, Paris, 2020.

Cet ouvrage a bénéficié du soutien des Programmes d'aide à la publication de l'Institut français. This work received support from the Institut Français through its publication program.

English translation copyright 2023 by the Regents of the University of Minnesota

All rights reserved. No part of this publication may be reproduced, stored in a retrieval system, or transmitted, in any form or by any means, electronic, mechanical, photocopying, recording, or otherwise, without the prior written permission of the publisher.

Published by the University of Minnesota Press
111 Third Avenue South, Suite 290
Minneapolis, MN 55401-2520
http://www.upress.umn.edu

ISBN 978-1-5179-1142-3 (hc)
ISBN 978-1-5179-1143-0 (pb)

A Cataloging-in-Publication record for this book is available from the Library of Congress.

Printed in the United States of America on acid-free paper

The University of Minnesota is an equal-opportunity educator and employer.

32 31 30 29 28 27 26 25 24 23   10 9 8 7 6 5 4 3 2 1

Contents

Translator's Introduction: A Shock to Think Together   vii
Thomas Lamarre

1. **The Question of Common Sense**   3
   Does Philosophy Confront Ignorance?   3
   The Defeat of Common Sense   9
   Problematizing Abstraction   17
   Civilizing Modernity?   25

2. **In the Grip of Bifurcation**   35
   The Bifurcation of Nature   35
   The Trick of Evil   43
   The Importance of Facts   48
   The Art of Conventions   56

3. **A Coherence to Be Created**   65
   We Require to Understand   65
   Dare to Speculate   70
   Whiteheadian Societies   79
   Heirs to Whitehead?   89

4. **What Can a Society Do?**   101
   Thinking through the Milieu   101
   Finding More   109
   Caring for Analogies   120
   Living Beings and Life   129

5. **A Metamorphic Universe** 139
Welding Imagination and Common Sense 139
In Praise of the Middle Voice 150
Tentacular Affects 159
Living in the Ruins 171

Notes 181

Bibliography 193

Translator's Introduction
A Shock to Think Together
THOMAS LAMARRE

The philosophy of Alfred North Whitehead has experienced a rather astonishing surge of attention in recent years. The surge has occurred especially in domains of inquiry related to media studies and science and technology studies, in which questions about culture and society are as important as the exploration of sciences, technologies, or media *tout court*. As Isabelle Stengers emphasizes in *Making Sense in Common*, Whitehead's philosophy is now engaged in dialogues and issues it never envisioned: antiracist struggles, Indigenous movements, climate-change activism, animal-rights advocacy. Stengers herself poses the challenge directly and lucidly: Does Whitehead's philosophy offer anything of relevance for the contemporary juncture? Her answer is a (prolonged) yes, and no one is better situated and equipped to address this challenge than Stengers. Not only has her work played a major role in generating the current surge of attention to Whitehead's philosophy, notably with *Thinking with Whitehead* (2002; English translation, 2011), which carefully yet boldly lays out a conceptual framework for understanding the genuine novelty of his manner of thinking, but it is also in her character as a thinker not to avoid troubled waters. The present book is not an exception: Stengers ventures here into unexpected domains of inquiry with Whitehead, charting a course across treacherous seas.

In this book even more than her previous work, Stengers brings Whitehead's philosophy into relation with the world of activist struggles. Her discussion begins with a consideration of how certain people have been barred from participation in scientific knowledge, or rather how certain practices of formulating questions have been ruled out, disavowed under the rubric of "common sense." This disavowal of the agency of certain people ("commoners" or nonexperts) or even entire peoples (Indigenous peoples, for instance) is closely connected to another kind of disavowal: the disavowal of the agency of nonhuman beings within certain forms of scientific practice. Although these two kinds of disavowal cannot and should not be conflated, it is not surprising that struggles against them are becoming increasingly interconnected in the contemporary world, finding common cause, so to speak. As Stengers shows, the question of the status of nonhuman beings within the sciences is not just pertinent to contemporary social and political activism. It may be of the utmost importance.

The question of how to avow the agency of nonhuman beings grips Stengers. She is exceedingly knowledgeable about a broad range of sciences such as chemistry, physics, biology, and ecology. But scientific knowledge for her is not simply a matter of facts about nonhuman beings, such as molecules, rocks, forests, animals, and the earth itself. What hold her interest are the ways in which (and the extent to which) nonhuman beings contribute to the production of human knowledge about them. Some scientific practices presume yet disavow the participation of nonhuman beings in knowledge about them. Indeed, the institution of Science has tended to encourage such practices, for political and financial reasons as much as epistemological ones. But scientific practice does not (and probably cannot) eradicate the agency of nonhuman beings. And it is here that Whitehead's philosophy has never been more relevant. Whitehead offers Stengers a meticulously systematic approach to thinking through the dynamic contributions of "actual entities" to the formation of "societies" at every gradation and order of complexity.

At the same time, Stengers remains keenly aware of peoples who have been forcibly and violently excluded from definitions of humanity established by modern Western humanism, particularly Indigenous peoples. What on earth has Whitehead to contribute to such questions? After all, it is not especially useful or desirable to make a wholesale equivalence between the marginalization of human beings and the marginalization of nonhuman beings, even if their struggles are frequently entangled. What kinds of relay are today possible between these two sites of contestation?

Stengers initially approaches this question from the standpoint of knowledge, questioning the status of different forms of knowledge and different ways of producing facts and speaking truth—"modes of abstraction." Here, too, her standpoint is above all that of activism, which allows her to think via struggles arising between different modes of abstraction. Yet her take on such struggles is idiosyncratic and sometimes unsettling, for she refuses to mobilize readers.[1] She rejects any form of mobilization that would encourage readers to accept the triumph of one side over the another. In some instances, such as the struggle against genetically modified organisms (GMOs) for agricultural use in Europe, she does not hesitate to take scientists to task, and one knows where her sympathies lie. Yet it is one of the hallmarks of her work that it shuns thinking in terms of victors and victories. She deliberately prevents her readers from concluding that scientists are entirely wrong while activists are in the right. She does not allow her readers to identify with any one cause. She refuses the comforting reassurance afforded by siding with either the victors or vanquished. This refusal of mobilization is Stengers's way of activating readers. Her tactic is, in effect, to combine activation and demobilization. The result is a restless, prickly style that may rub some readers the wrong way, for if no one may be declared the victor, neither is anyone to be considered just a victim.

It is to reactivate our thinking about contemporary struggles that Stengers introduces terms and concepts that some readers are likely to find unfashionable, dubious, questionable, and even objectionable, and as Stengers herself would add, for good reason. This is especially true of the key concept of this book, common sense. The very notion of common sense, because so closely associated with consensus, feels outmoded, surpassed, and ultimately exhausted. What can Stengers be thinking when she proposes reactivating, of all things, common sense? The same sorts of questions arise around other key terms in the book, in particular "civilization" and (Western) "modernity." These terms feel not merely outmoded, but potentially objectionable in light of the histories of colonial violence associated with them. Although Stengers does not propose to reactivate the paradigms of civilization or Western modernity as such, neither does she reject them outright. For Stengers, terms such as "common sense," "civilization," and "Western modernity" are mobilizing terms. They spur readers to take sides without thinking. Her tactic, then, is to activate thinking by pushing readers to demobilize their responses to such mobilizing terms. Without sustained and systematic philosophical engagement, such a tactic of provocation might appear simply abrasive, or even abusive. It is useful, then, to provide some sense of her philosophical method at the outset.

Prior to her deeper engagement with Whitehead, in addition to a series of monographs coauthored with thinkers with areas of expertise outside her principal fields of chemistry and philosophy, such as physics (Ilya Prigogine), German philosophy of nature (Judith Schlanger), and psychiatry (Léon Chertok), Stengers penned a series of interventions into the history and philosophy of sciences, gathered in *The Invention of Modern Sciences* (1993) and the seven volumes of *Cosmopolitics* (1997), the latter of which was subsequently published in two volumes. In these essays, Stengers draws a good deal on the philosophies of Gilles Deleuze and Gottfried Leibniz. Yet, where one might expect fusion, tacit agreement, or harmonious synthesis of the two philosophers, Stengers's approach is characterized by an unflinching respect for the positions of both philosophers, which leads her to develop a sharp contrast between them. Thus, when Stengers refers to her new diplomatic criteria for sciences as "the 'Leibnizian constraint' according to which philosophy should not have as its ideal the 'reversal of established sentiment,'" she immediately notes Deleuze's disapproval of Leibniz's stance.[2]

This passage is revelatory of Stengers's method in a number of ways. First, it anticipates her focus on common sense in this book in relation to Whitehead, which she had previously articulated via Leibniz's respect for established sentiment. Significantly, Stengers provides a footnote for Leibniz's remarks, and yet, where one might expect a reference to Leibniz's text, she cites Whitehead: "You may polish up commonsense, you may contradict in detail, you may surprise it. But ultimately your whole task is to satisfy it."[3] Her engagement with common sense and Whitehead, then, is announced early in her career, and clearly. In this respect, *Making Sense in Common* is a summation of Stengers's work to date. It also marks a turning point. In her preface to the original French edition of the book, she expresses her desire to take on the challenge of a new mode of address to readers.[4] In this book, she aims to speak not only to the converted (those who are likely to agree with her or already in the know) but also to readers who may well disagree with her. As a result, *Making Sense in Common* is the most accessible of Stengers's books, at once an introduction to and summation of her thought. It does not, however, vulgarize it or dumb it down. Instead, while she strives for clarity of expression, she places even greater weight on developing the contrast between opposing sides of any struggle, which makes her diplomatic mission all the more difficult. *Making Sense in Common* is at once the most contrastive and the most diplomatic of her works to date.

Second, her source for the Whitehead citation is *The Aims of Education*, which underscores how Stengers's concern for knowledge production extends pragmatically (diplomatically) into education. Her aim for education is far from Socratic, however. She devotes the first chapter of *Making Sense in Common* to distinguishing her approach from that of Socrates. Her position as a philosopher is not that of "he who knows better precisely because he knows he does not know, and thus may teach." Stengers avoids this movement of *mise-en-abîme* in which knowledge turns into a dizzying and disorienting hall of mirrors around the *aporia* of knowing. Education for her never hinges on teaching those who do not know that they do not know. It is about learning by way of what matters to us, that which we do know. The philosopher for Stengers is not a teacher, authority, or specialist. The philosopher is one who cares enough about something to learn more about it with others. Her task is to *find more*. This often means the philosopher must engage in diplomacy at the same time: it impossible to find more with others if one does not also keep the peace with them.

Third, when Stengers dramatizes the contrast between Deleuze and Leibniz with respect to common sense, her contrast does not imply a rejection of Deleuze in favor of Leibniz. Indeed, her philosophy owes so much to Deleuze that her fellow traveler Bruno Latour refers to him as her only true mentor.[5] Perhaps for this reason, disagreements with Deleuze pepper her work. Common sense is but one example. Another prime example elsewhere in her work is her reconsideration of Deleuze and Félix Guattari's distinction between the sedentary and the nomadic in *A Thousand Plateaus*. Just as Stengers seems here to take the side of common sense, so she apparently sides with the sedentary over the nomadic.[6] To understand her tactics, it is important to recall that she never thinks in terms of victors or of victories in which one side is right and the other wrong. In this, her disagreement with Deleuze and Guattari is ultimately true to their own method. Even if they seem to revel in the ways in which the nomadic uproots sedentary habits, the former is not a category for them, but a tendency. It is impossible to think or to live the nomadic without the sedentary, or vice versa. The nomadic cannot be triumphantly celebrated as if in victory over the sedentary. At stake for Deleuze and Guattari is a logic of disjunctive synthesis and a practice of assemblage. Similarly, when Stengers seems to side with the sedentary, the nomadic remains in play. By the same token, un-common sense is in play in common sense. In effect, when Stengers disagrees with Deleuze, it is to introduce and develop a contrast in accordance with what might be called somewhat awkwardly "contrastive synthesis." Thus, when she insists on the importance of not shocking common sense, she remains highly attuned to the importance Deleuze ascribes to the nonsensuous and nonconscious that shocks us to thought. The question then is: where and how will Stengers situate that sort of shock?

Fourth, when Stengers insists that, above all, one must not shock common sense, it is important to note that she is situating shock on the same level as common sense. This introduces a complex parity between what shocks and what is held in common. Consequently, common sense or the "in common" is always situated in relation to collective endeavors and struggles. What we hold in common is not placed in opposition to what shocks us to thought, to nonsensuous and nonconscious activation. Her stance, then, implies that a "shock to thinking in common" or a "shock to think together" is already taking place.

When Deleuze was writing *Difference and Repetition* or when Deleuze and Guattari were writing *A Thousand Plateaus*, it may have made tactical sense to stress the shock to thought, and even to sensationalize it. For her part, Stengers does not disavow what shocks us to thought: it is because the shock to thought is never distributed evenly or equitably that is has the power to transform the world. Indeed, Stengers is not loath to resort to shock. For instance, in her reprise of Donna Haraway's account of training with the dog Cayenne, Stengers remarks: "A zigzag may be generated out of the shock that the term 'trust' may arouse in some people." Today, however, in an era confronted with fake news, uprisings, and climate change, Stengers believes that our tactics must change. Amid constant shocks to thought, we also need ways to buffer or protect ourselves if we are to "make sense in common." We need to avow our vulnerability and understand what makes us vulnerable. Making sense in common arises in the interval between the Deleuzean shock to thought and the Whiteheadian lure to feeling. Making sense in common arises at the same level as the shock to think together.

Fifth, Stengers makes frequent reference to the defeat or undoing of common sense. The defeat of common sense means that neither scientists nor humanists take it seriously; they reject it to mobilize among themselves. Politicians and mass media do not take it seriously either; they mobilize it to accrue attention and profit. For Stengers, then, common sense is a mobilizing term in a profoundly practical way: so much is mobilized against it; its defeat has been triumphantly declared by the victors to further their cause. Defeat, then, does not amount to demobilization. Defeat for Stengers is a sign of total mobilization. Her aim is consequently to demobilize common sense, which does not mean rehabilitating or redeeming it. On the contrary, Stengers herself undoes the notion of common sense, discovering in it something of a contrastive synthesis: "sense" implicates diverse and heterogeneous orientations, while "common" implies unity. Equally concerning for Stengers are the temporality and historicity of this implied unity. She renounces placing unity in the past. While she stands against the developmentalism associated with capitalism—that is, economic development for its own sake—Stengers is nonetheless resolutely non-Romantic, without nostalgia for a retrospectively idealized past. Stengers turns instead to a series of pragmatic ways of making sense in common and aiming at a relevant future, among them the Quaker meeting, the palaver, and Latour's guidelines for a renewed agora, all of which evoke a past in the immediate present. Stengers thus makes clear that the unity of the common belongs to what Whitehead calls the relation between the immediate present and the relevant future.[7] Unlike consensus, then, which implies an agreement to be reached and held, the unity of the common aims at the relevant future. Making sense in common is an ongoing process, an aim instead of a goal.

Sixth and finally, Stengers continually returns to a passage from Whitehead that becomes a catchphrase for her project: the "welding" of common sense and imagination. The metallurgical operation of welding affords a fitting image of Stengers's method. When two pieces of metal are welded, they are operationally situated on the same level, even if it is an edge joint or tee joint or lap joint. The two pieces remain distinct, yet now belong together in an assemblage, and one in which the joint too is tangible. Contrast becomes palpable while introducing another dimension. Such welding is precisely what Stengers aims at in her approach to common sense and imagination, which is also to say, the pragmatic and the speculative dimensions of her thought. Thus, she draws on another of Whitehead's fecund turns of phrase: "common sense brooding." Common sense, however pragmatic, has something speculative about it. As for imagination, although Stengers does not speak to it as such, she puts her sense of it into action. Imagination, however prone to speculation, introduces a pragmatic turn when it allows for a mode of generalization, which implies a mode of abstraction that is neither inductive nor deductive, but imaginative. This mode of imaginative generalization that is produced when imagination is welded with common sense is called a "generative apparatus."

In writing this book, Stengers constantly consulted the English editions of Whitehead alongside the existing French translations. Often, she adds footnotes explaining how and why she modified the French translation, many of which she deemed unnecessary to reproduce in the English translation. Needless to say, as Whiteheadian turns of phrase like "common sense brooding" suggest, translation of Whitehead into French is no easy matter. "Brooding," for instance, becomes *rumination* in French. I might equally well have rendered the "brooding of common sense" as the "rumination of common sense." This shift in tonality introduces a challenge. When the word *rumination* subsequently occurs in the French text, is it still "brooding," or is it now "rumination"? This may seem a trivial matter, but given the highly expressive, even poetic nature of Whitehead's prose, in combination with its meticulous array of prepositions and conjunctions to highlight conceptual operations, it turns out to be of the utmost importance. I opted to retain some terms, such as "brooding," to give a sense of how Whitehead's prose affected Stengers's thinking, for there is a way in which Whiteheadian turns of phrase seep into her turn of thought. Indeed, Stengers sometimes adopts mannerisms from Whitehead. For instance, she often begins sentences with "and," an unusual gesture in academic French, a perturbance she wishes to prolong. But the same gesture in English no longer perturbs. What is more, when Whitehead starts sentences with "and," his gesture is inseparable from a system of thought comprising a use of conjunctions, prepositions, and relative clauses that is impossible to sustain in translation, and surely undesirable, for the result would be an overcoding of Stengers's style. And so, I adopted other turns of phrase, and even forms of repetition, that seemed to me to convey the tone of Stengers instead of trying to imitate Whitehead. In many instances, keeping to the spirit of her thought resulted in a radical departure from literal translation. In fact, Stengers herself opted to alter and expand some passages with me for this edition, not only for the sake of clarity and precision, but also to convey something of the tonality of thinking with Whitehead through a more contemporary English.

Our dialogues over terms and passages raise a deeper question of translation: how to convey how Stengers seeps into Whitehead? In the original text, because she is translating and modifying some of the existing translations of Whitehead into French, her thinking permeates Whitehead's. When passages from Whitehead are returned to the original English, the effect can be jarring, and not only stylistically, but conceptually. A chasm opens between his English and her more contemporary prose in English, where the two styles felt more entangled in the original French. This divide risks making her approach feel like a form of exegesis of passages from Whitehead, when nothing could be further from Stengers's undertaking. She refers to her approach as a way of relaying Whitehead, of taking up the baton from him. She also remarks that Whitehead's language is not our own, by which she means that it belongs to a time and place implying particular problematics. It is this "relay" of problematics (or the ingression of eternal objects) that matters to her.

To give some sense of how Stengers affects Whitehead, I tried, whenever possible, to emphasize her turns of phrase that alter how we read Whitehead. One example is her use of *fait*, both as "fact" and as "done." This emphasis on the production of facts is in keeping with Whitehead, and the effect is reinforced with the continual use of compound verbs with *faire*, such as *faire sentir* and *faire prévaloir* and *faire compter*. The doing implied in such compounds may be somewhat literally translated as "make feel" and "make prevail" and "make count." Not only does such a translation eventually prove awkward and feel tendentious if overused, but it also fails to capture the subtle turn Stengers introduces, which is akin to James's expression "in the making" or Whitehead's "to be created." *Faire sentir* approaches "feeling in the making." Another example is Stengers's use of language related to hands and tactility or hapticity. Like Whitehead, she disassociates feeling from perception and personalized emotion, and tactile terminology highlights the force of a nonpersonal and nonconscious feeling. Thus, in addition to deploying terms related to grip and grasp, she highlights echoes of Whiteheadian prehension in verbs like *apprendre* (apprehending or learning) and *comprendre* (comprehending). It would be too much, of course, always to tack on "in the making" or "make" in every instance of *faire*, or to render *apprendre* as "apprehending" when "learning" is more appropriate. Yet another example is Stengers's use of the language of taste, tasting, and savoring, which insinuates an unsuspected yet pertinent dimension of experience into Whitehead's articulations. In all instances, I aimed for a judicious introduction of turns of phrase that might express something of the stylistic and conceptual entanglement of Stengers and Whitehead in the original. Indeed, I found myself thinking and practicing translation in the manner of generative apparatus, in Stengers's terms. Translating, like Stengers's relaying of Whitehead, can be made into a mode of making sense in common. I am grateful for Stengers's participation in that process.

The remarkable achievement of *Making Sense in Common* is to show us how to take what we know and what shocks us to think together, and to weld them into a generative apparatus for making sense in common, through the use of contrastive synthesis. In a time in which it is impossible to deny that living on a damaged planet is upon us, Stengers, with Whitehead, teaches us how to live in the ruins by imagining a future relevant to what we know.

Making Sense in Common

{ 1 }

The Question of Common Sense

The citizens of Athens never imagined that Socrates was going to transform them into his patsies, allowing him to justify the adventure of this strange undertaking that has been called philosophy ever since. Nor did I imagine that Alfred North Whitehead, such a radically atypical philosopher, would resort to such an utterly typical remark about the origins of philosophy as this: "Socrates spent his life in analysing the current presuppositions of the Athenian world. He explicitly recognized that his philosophy was an attitude in the face of ignorance."[1] This sort of commonplace remark is not in itself noteworthy. It is common enough in Whitehead's texts, and each reader must decide on their own whether to find a reason to ignore it and to move on to the next phrase with a smile, or to stop and puzzle over it. I have opted to grant it the power to put me to work.

Does Philosophy Confront Ignorance?

When I imagine how Socrates's peculiar questions caught the citizens of Athens off guard, what first come to mind are the posters that caught my attention in one of the corridors of the European Commission building, which houses the offices of civil servants tasked with questions about "science and society." The posters reproduced the results of public opinion polls dealing with what European citizens think about science. In light of the absurdity of the opinions expressed, the results seemed to have been posted as a reminder to civil servants about what sort of attitude would be suitable when dealing with a band of unabashedly ignorant people, whom one must pretend to respect, but above all, must manage—all for their own good, of course.

As everyone knows, pollsters count on the agreement of those whom they kindly call "average citizens" to provide responses to questions without wondering what sort of trick is being played on them, even though they have never before had the opportunity to engage with such issues. Pollsters show no scruples about relying on the weakness of those whom they trap, which simply makes them crooks. But of course, Socrates was not a crook. He went to great lengths to make the citizens whom he questioned see the ignorance evidenced in their answers. This is what Whitehead calls an attitude, and the term "attitude" takes on a good deal of weight. The attitude of Socrates is itself a philosophical theme. There are as many possible Socrateses as there are readings of his attitude toward ignorance, and as many ways of facing the beginnings of philosophy.

One Socrates is a master of aporia, claiming not to have any answer himself but seeking only to make his interlocutors confront the difficulty, possibly insurmountable, of formulating the answer. He is the one who knows himself ignorant. There is another Socrates, master of Plato, for whom aporia is a form of propaedeutics, preparing citizens to welcome a knowledge that transcends the divergent answers they have proposed. He is one who invents philosophy to pacify disagreements, giving to the city his orientation toward what is truly good, just, and beautiful, above and beyond illusions. Yet the historical Socrates was condemned for poisoning the public peace, for instilling the poison of doubt, and Wittgenstein may well have ratified this condemnation, Wittgenstein the anti-Socrates who passed his life posing questions not so much of ordinary citizens as of his fellow philosophers, who stood accused of spreading the disease of false problems.

There can be many possible Socrateses, yet those whom he addresses are always construed as ignorant citizens. Asked to define truth, justice, or courage, they offer cases and examples that, as Socrates shows easily enough, lead to divergent definitions. A veritable stingray, Socrates sparks them awake in an attempt (it is said) to share with them his only privilege, of knowing he does not know. It could also be said that he stuns them. He leaves them stupefied, convinced of their incapacity to know what they are saying, and ready to leave it to the philosopher to guide them. Or else he confronts them with a challenge that is also a trap. If the meaning of words indeed depends on making reference to circumstances or a particular language game, the citizens of Athens were not ignorant. They knew everything there was to know.

What would Whitehead's attitude in the streets of Athens be? In *Modes of Thought*, he praises the practice of assemblage, to be taken up time and time again from era to era. He associates assemblage with what for him is the task of philosophy: "Philosophy can exclude nothing."[2] Assemblage changes everything. The different answers the philosopher gathers, however divergent and partial they may be, are not to be disqualified or reduced to attesting to a speaker's ignorance on the part of a Whiteheadian Socrates. They are part of an assemblage that puts the philosopher to work. What characterizes the assemblage is the *problematic*. The problematic is not a problem to be resolved, for even if an answer proves correct, it will continue to vie against others. The problematic implies a terrain to be shared under the aegis of perplexity activated by the philosopher. If Socrates had not positioned himself as an arbitrator, judging and excluding, he might have been able to make the divergence revealed by his requests for definition into a source of collective concern. He might have welcomed the perplexity he aroused, not as a symptom, but as a question, which he might have shared with what Whitehead calls common sense: "Common sense brooding over the aspects of existence hands [them] over to philosophy for elucidation into some coherence of understanding."[3]

Ignorance here shows an entirely different face. The citizens "awakened" by Socrates will not have to abandon their initial propositions as worthless. Socrates's questioning caught them off guard. They know that they have let themselves be surprised by an unusual question, and in this respect, their ignorance has been demonstrated. Even if they expressed them only partially, their propositions attest to a knowledge that need not for all that be cancelled out. Another Socrates must be imagined, one who needs the brooding of the citizens of Athens. This Socrates needs citizens who accept that there is no need to lend authority to the commonplace propositions bedecking their thought, but neither is it necessary to disavow those aspects of existence that matter to them. On the contrary, their brooding should activate the sense of importance that they attached to their answer, which related it to an aspect of existence. Belonging to existence itself, this aspect would be irreducible to what is habitually attributed to the "subjective" and relativized. So it is that Whitehead eschews using the divergence of answers to deny their value: "The philosophic attitude is a resolute attempt to enlarge the understanding of the scope of application of every notion which enters into our current thought. The philosophic attempt takes every word, and every phrase, in the verbal expression of thought, and asks, 'What does it mean?'"[4]

In *Modes of Thought*, Whitehead relentlessly activated words. His mode entailed reimmersing words in situations that were part of common experience without letting common experience define them. Instead, common experience takes on the power to engage words in a *speculative* adventure. Neither metaphoric nor literal in meaning, "speculative" here means dramatizing what goes without saying when we say something. It deploys what is presupposed and assumed by the most limpid and banal of statements, even to the point where we lose our bearings, so long as the statement is not reduced to *a* statement, but envisaged as *this* statement, always engaged in *this* situation, answering to *this* mode of engagement in the situation: "Philosophy begins in wonder. And, at the end, when philosophic thought has done its best, the wonder remains. There have been added, however, some grasp of the immensity of things, some purification of emotion by understanding."[5]

When philosophy has done its best, the sense of the word "wonder" has somehow changed. At first, as the philosopher encountered a discordant multiplicity of meanings demanding elucidation, "wonder" signaled perplexity. After she has endeavored to *comprehend,* what remains is closer to wonderment, for she has grasped something of the immensity presupposed and claimed by each aspect of existence. Philosophy does not respond to the brooding of common sense with procedures of selection and hierarchy to eradicate discrepancy. Nor will philosophy pacify common sense by assigning a meticulously bounded territory to each aspect of existence. Such solutions do not inspire wonder. They sadly accept limits that commit them to thinking under surveillance—the triumph of critique: "The strength of the critical school lies in the fact that the doctrine of evolution never entered, in any radical sense, into ancient scholarship. Thus there arises the presupposition of a fixed specification of the human mind; and the blueprint of this specification is the dictionary."[6]

If the citizens of Athens had been armed with a dictionary to provide a fixed answer, or to define the rules of good usage, or to trace the more or less arbitrary philological evolution of a meaning, then they might have answered Socrates's questions. But they would have done so as if at school. Such a mode would have shielded them from any perplexity and from any grasp of the immensity of things. The ability to respond to the Socratic question "what is . . . ?" with a definition of what we mean by "courage," or "the good," or "justice" produces a statement stripped to its bones, stripped of its sensuous flesh, soft and corruptible. The dream of the ideal dictionary is to extract only what proves resistant to critique, intent on what belongs to the human mind as such, independently of shifting and diverging sensibilities. Stripped of all reference to passions, appearances, and circumstances, such a mind would for Whitehead be nothing but an automaton, incapable of error perhaps, yet incapable of understanding as a result. What proves resistant to critique would be nothing more than dead abstractions, to be passively accepted because they do not arouse any stirring of thought or of imagination.

In contrast, to accept the doctrine of evolution in a radical manner is to accept a form of empiricism that embraces change as primordial. Evolution does not produce species that remain fixed. The designation of a species, including the species called human, is based on the relative stability of a handful of traits permitting characterization and classification. It does not set limits on what the individual realities thus identified are capable of. Accepting evolution means agreeing to abandon the idea that thought needs fixed references to avoid confusion and arbitrariness. Evolution strips critique of its power when it strives to hold us accountable by demanding guarantees and discounting what it calls beliefs. For Whitehead, there is no stable definition of common sense any more than there is a fixed identity to the human species. There is no way to define a common sense that would allow us to ground consensus, nor one that we would have to resist.

The aim here is not to define common sense, much less to envisage some sort of philosophy of common sense, and still less to make common sense an attribute of the human. Whitehead made common sense a constraint for philosophy. Common sense is a constraint for philosophy because it is the task of philosophy, as Whitehead understands it, to refuse the kind of freedom specialists claim when they rule out or exclude what is incompatible with their presuppositions, even taking pride in scandalizing common sense. To respect a constraint, however, is not to respect a limit. It is to refuse what comes easy. What comes easy would be to accept variability with nothing at stake, to assume that the question "what does that mean?" is entirely arbitrary, depending on the moment.

Evolution for Whitehead is not a matter of progress toward the human. Neither is it an arbitrary history, something merely observed. His key word is adventure. The calling of philosophy is to consent to adventure, which means participating in it. The task of philosophy thus requires, as Whitehead puts it, "a welding of imagination and common sense."[7]

The originality of Whitehead as a philosopher, then, comes of his speculative relation to common sense. Common sense is not only a constraint but also a wager. If it is to be welded to imagination, common sense must be capable of brooding. "Brooding" implies not being taken in, not docilely agreeing to disqualify what matters to it. Common sense cannot be reduced to what philosophers discuss or what they define. It cannot be reduced to playing a role in their thought, whether authority or patsy. The possibility of welding, which implies a genuinely metallurgical operation, is speculative. Its wager is adventure instead of progress. The possibility of welding implies that philosophy does not ultimately have to bring a satisfactory answer to the brooding of common sense. It has to nourish what makes for brooding. Such is Whitehead's attitude toward the ignorance that Socrates forced the inhabitants of Athens into admitting. Ignorance is our common lot as we face the immensity of things. The question, however, is not to know that one does not know. That is just another way of discrediting common sense. It is a matter of daring to imagine, contrary to the assurances of specialized forms of knowledge, that what mutely insists and makes us brood expresses a certain grasp of the immensity of things—even if we don't quite know how to put it into words.

The Defeat of Common Sense

The task Whitehead proposed for philosophy, welding common sense and imagination, came in direct response to what he considered to be the mortal, even fatal weakness of the modern world: discrediting common sense, especially through specialized theories that revel in persuading common sense to give way to the authority of "those who know." The task of philosophy, then, might be said to be one of *reactivating* common sense, since we are familiar only with common sense as it has been defeated.

There are many ways to tell the story of its defeat. Some ways are political. Gilles Deleuze, for instance, proposed that between the left and the right there existed a difference not of sensibility or of priority, but of nature. The left has a vital need for people to think, while the right requires them to comply with a self-evident order of things, to agree to formulations of problems coming from elsewhere. To defeat common sense, then, is to render brooding powerless. Brooding is stripped of all capacity to object to the received order of things. Brooding is reduced to a plaintive imaginary, dreaming, if at all, of a world in which the so-called people would not be so egotistical, irresponsible, and so easy to influence.

Leibniz wrote that life is everywhere but not everything is living.[8] Similarly, the political may be found everywhere, but not everything can be reduced to a "properly" political register, at least not until we have the means to pose the question in this way. The political register must first become capable of taking on a proposition like Whitehead's, capable of translating in its own terms what is meant by welding imagination and common sense. There is no way to cut corners here. Many will find Whitehead's proposition meaningless, or they will immediately hear in it a call to make imagination comply with conditions imposed on it by a common sense that must be respected.

How, then, to respond to a physicist who raises these sorts of objection: "If we had respected common sense, neither Einsteinian relativity nor quantum mechanics would have seen the light of day! And Galileo! How could he have convinced people that the Earth moves around the sun without their being aware of it?" This latter argument, it should be noted, is not a sound move. After all, Galileo published *Dialogues on the Two Chief World Systems* in Italian and not in Latin, and he had Sagredo, who presents himself as an enlightened amateur, play the role of arbitrator in his polemic with Simplicio, who represented his adversaries. Galileo thus placed himself on the side of common sense (Sagredo) against authority (Simplicio). The fact remains that everything changes at the start of the twentieth century.[9] Physics then takes on greater authority, and its authority is subsequently defined as revolutionary, as destroying the certainty of what is now, correlatively, characterized as common sense.

It is worth highlighting how strange this situation is. For physics to define itself against common sense, it first had to define common sense to suit its purposes. Through a strange sleight of hand, the certainties physics attributes to common sense are precisely those that characterized the so-called "mechanistic" view of the world prior to the revolution in physics. In other words, "people" had to be convinced that they "naturally" think in terms that in fact belong to what is now defined as classical physics. They had to be convinced, for instance, that they had always adhered to the idea of a physical reality consisting of particles in motion, each one characterized by a well-defined position and velocity. A great deal of instruction is needed to make what proves scandalous to physicists feel like a scandal for all of us, such that we indeed feel consternation when the physicists declare, "we can no longer claim a direct access to reality!"

A scene typical of this academic squabble: an enraged physicist challenges a critical thinker. The latter may be a philosopher, a sociologist, or a specialist of science studies or cultural studies. It does not matter to the physicist, who is not interested in such distinctions. What counts for him is what he has perceived: his interlocutor was likely to introduce the shadow of a doubt into the mind of the public as to whether the laws of physics really and truly describe reality. He cannot condemn this new Socrates to drink hemlock, but what he proposes has nothing amicable about it: "If you do not believe in the laws of physics, throw yourself out the window!" Especially distressing is that this ploy seems to work. The almost ritualistic character of this challenge—I can no longer count the number of times I have heard it, read it, or experienced it—testifies to the fact that physicists resort to it with impunity, apparently without fearing that someone might reply in surprise: "Do you mean to say that, before Galileo and his law of falling bodies, people remained blissfully ignorant of the difference between doors and windows?" Their adversaries, however, when they turn to deconstructing scientific objectivity, do not prove equal to the challenge. To counter the example of airplanes flying as proof of the objective existence of the laws of physics, one critical thinker even evoked the potential efficacy of prayer. He seemingly did not fear the shocked objection his argument could raise: a believer, however fervent, might reflect and then object that he would never choose an airline that replaced technicians with prayer circles; in other words, his trust in the safety of flight is not in the objective laws of aerodynamics when he boards a plane, but in the laborious work of airplane maintenance, pilot training, alertness of air controllers, and so forth.

The impunity they feel in rejecting common sense makes idiots of "those who know." The science wars took place in a land without people. A de facto situation defined the clash of the warriors: the defeat of common sense. And the defeat did not take place on any battlefield whatsoever. Strictly speaking, a battle never took place, because common sense has neither weapons nor a specific cause to defend when it comes to what was really of interest to scientists, in this case not the fact that heavy bodies fall, but the manner in which it is possible to define the variation of their velocity during their fall. Instead of a defeat, it was more of a *dé-fait,* a de-facting or undoing, a dissolution of the capacity to object, which also means the loss of what makes brooding possible.

Voicelessly perhaps, and without necessarily deploying a contradictory discourse, to brood is to refuse to lose confidence in the value of experience, even if experience is difficult to put into words, even if it is placed in a position of difficulty by a theory aiming to discredit it. What is undone is the possibility of a relation of hesitation with respect to theoretical claims. Hesitation differs from a relationship of distrust with respect to theories as such. Hesitation does not imply distrust, but the capacity for a positive appreciation of what a theory proposes, without for all that allowing the theory to deny what it leaves out or reduces to appearances. The corollary to the undoing of common sense, then, is the ignorant arrogance evident in so many theories. Theory without relation to common sense is like a boat whose rudder has lost contact with the water, spinning in circles, unable to maintain its course, which is to say, unable to avoid stupidity.

While the mass media have lost interest in the science wars, the war goes on. The science wars, which spread their stupidity with impunity, are part of a past that has not passed. They surely would have captured Whitehead's attention. But the clash between scientists and critical thinkers would not have surprised him. For Whitehead, the transformation of philosophy into a specialization is evident in the incapacity of critical thinkers, from Hume and Kant, to "speak well" about what the sciences celebrate as triumphs. He notes: "The question 'what do we know?' has been transformed into the question 'what can we know?'"[10] This transformation sets up philosophy to judge human experience, to gauge what humans, including scientists, "know," even if they do not know how to put it into words. The broodings of common sense have been reduced to a form of confusion deemed inherent to questions that exceed the limits of what we may legitimately claim to know, of what we are entitled to know. If someone broods, it is because they are wandering in the labyrinth of their beliefs. As for those scientists who claim to have access to reality, criticism shows tolerance for their naïve realism. In response, Whitehead writes, "scientific faith has risen to the occasion, and has tacitly removed the philosophic mountain."[11]

Before he became a protagonist in the science wars, Steven Weinberg neatly expressed the amused scorn the philosophic mountain inspired in him:

> Ludwig Wittgenstein, denying even the possibility of explaining any fact on the basis of any other fact, warned that "at the basis of the whole modern view of the world lies the illusion that the so-called laws of nature are the explanations of natural phenomena." Such warnings leave me cold. To tell a physicist that the laws of nature are not explanations of natural phenomena is like telling a tiger stalking prey that all flesh is grass. The fact that we scientists do not know how to state in a way that philosophers would approve what it is that we are doing in searching for scientific explanations does not mean that we are not doing something worthwhile. We could use help from professional

philosophers in understanding what it is that we are doing, but with or without their help we shall keep at it.[12]

Weinberg waxes ironic without displaying any anger. Scientists and the majority of philosophers have, in fact, always been in agreement on the essential: sciences attest to human rationality. Everything changed when criticism ceased to be respectful, when certain critical thinkers made explicit the consequences of their deconstructive undertaking: " 'Nature' does not play an active part in your agreement about the order you decipher from it, thus your agreement is merely human: you scientists come to an agreement among yourselves, and you do so in a manner that differs not the least from what holds for any other human agreement." Exeunt rationality. Yet rationality is not replaced by a genuine interest in the multiple ways in which humans are liable to come to an agreement. When the critic affirms that science is "nothing but a practice like any others," it is the "nothing but" that counts. The critic claims that agreement about facts have the same basis as any practical agreement. Reaching a verdict on facts is simply a result of relations of force.

What the science wars made apparent is that common sense is now nothing other than public opinion, subject to influence, readily seduced by the critics' proposition. Rather than treating them with irony as they once did, scientists accused critics of promoting irrationality. Suddenly, relativism was turned into a monstrous threat, the idea that all practices are equal in value, even those of quacks and primitive peoples! Apparently, the loss of authority of the laws of nature was taken as synonymous with an outbreak of arbitrariness. Indeed, scientists seemed always to return to these laws, to theoretical physics, even if it meant adding Darwinian evolution to designate creationists as the real enemies. Without these laws, nothing would hold back the most irrational passions. All would be permitted, each person with their truth, even flat-earth advocates.

It is now accepted that "people," descendants of the citizens of Athens, are no longer capable of questioning, of brooding. What Whitehead dubbed "common sense" is no longer something to be reckoned with, or to be counted on. It is now taken for granted that the man in the streets or the country dweller is mistaken. Indeed, this is the only thing specialists agree on unanimously. Worse, people are seen as susceptible to more than being mistaken; they are accused of being ready to follow the first demagogue to come along. They are thus in need of shepherds. Since people are disposed to believe in anything and everything, any argument is permissible for the sake of denouncing bad shepherds.

Today this caricature has taken on life and entered in the political arena, or more precisely, the electoral arena, the only arena in which people are granted the right to make their voice heard. The defeated common sense, rendered incapable of brooding or even of taking offense, the perfect strawman, has rather suddenly given birth to the monster scientists evoked against those critics they treated as relativists. Of course, faced with the shameless proliferation of "fake news" and "alternative truths," faced with the resolute indifference to or virulent defiance of well-established facts, one might be tempted to settle accounts from twenty-five years ago: "As you can see, relativism, in its attack on the authority of facts that should have established agreement, has indeed given rise to irrationality. We were right, and you have let a terrible genie out of the bottle."

One may interpret what has happened from different points of view, obvious ones such as the role of social networks and disinformation campaigns of all sorts, but also the lived experience of seeing the progress in which we were asked to believe vanish from the horizon. But it seems to me that the daunting novelty of the event is overlooked when we resort to an image of the public that confirms its fundamental irrationality, for the sake of lending credence to the thesis that we must trust in those who know, or else anything goes. It strikes me that those who revel in fake news and other alternative truths seem less to manifest a blind credulity than a dark unwillingness to understand anything in order to take revenge on those who know. Who would have believed that Donald Trump, during his electoral campaign, would declare with impunity, "I could stand in the middle of Fifth Avenue and shoot somebody, and I wouldn't lose any voters"? His voters surely understood that Trump was offering them a splendid opportunity to further scandalize the intellectual elite.

What if such blind hate against this elite "who knows" was related to the cultural and political disaster which I am calling the defeat of common sense? The words of Bertolt Brecht spring to mind:

"We often speak of the violence of a river overflowing but less of the violence of the banks that confine it."[13] He was surely thinking of revolutionary violence and people in the streets, not of these crowds who so passionately applaud the most outrageous and openly absurd of statements, discovering that they count for something because they inspire fear in those who know. While the overflowing rivers tend to differ, the violence of the banks is similar. Our shepherds were wrong to believe that they could with impunity make those whose confidence they solicit feel they have absolutely no reason to take offense when promises are not kept. People cannot be humiliated with impunity.

We must not let the response of Trump supporters (and others who reject as falsehood and conspiracy anything that requires thought of them) obscure the existence of radically different responses to the mirage of progress. Other responses arise that produce a recalcitrant thinking, putting to the test reasons supposed to persuade. Increasingly numerous among the descendants of Athenian citizens are women and men who no longer brood yet do not turn into fools as a result. They know what they resist when they take to protesting in the streets, or mowing down contaminated fields (where GMO are grown), or camping in "zones to be defended," or going on strike against "objective" managerial redefinitions that destroy the meaning of their jobs. Or else, they experiment with alternative therapeutic practices regardless of data-based expertise. In all such cases, they call into question the definition of the object on which the experts base their authority, not with the violence of those who refuse to listen to reason, but with the intelligence of those who have learned to resist its authority. They contest an authority based on ignorance or rejection of what they know, what they feel matters. They insist on taking multiple interdependencies into account, entanglements of humans and nonhumans, which the experts insist on forgetting, and which they too had become accustomed to forgetting all too often.

To think in terms of the defeat of common sense is to resist the temptation to say that the first group, those who destroy the riverbanks, who have decided not to listen to reason anymore and to remain deaf to anything that would make them doubt, are in the wrong, while the second group, the activists, are in the right. Both are products of the same disaster. We may, however, side with the manner in which the second group gives meaning to this disaster, in contrast with the manner of the first group. That is what I propose to do, while keeping in mind, as a philosopher, that we are no longer in the streets of Athens. Those who struggle will certainly not ask philosophy to help them shed light on what makes them struggle. It was not philosophers who opened the pathway for activists. Instead, philosophers may benefit from what the men and women who engage in activism have to teach them.

Philosophers can learn what it might mean politically to weld imagination and common sense. Knowledge that activates does so in the mode of resistance and struggle, which lends it importance and relevance, but which may equally well make it vulnerable, making them lay claim, for instance, to the legitimacy of the concrete in opposition to the usurpation of the abstract. The struggle to loosen the grip of definitions authorizing objective knowledge is justified when engaging the enemy. But the genuine force of activists comes of taking on the challenge of this question: how not to resemble the enemy? Implied in this question is another: how not to attribute the power of mobilization to a truth that we declare concrete, thus unifying everyone like good little soldiers? In other words, how not to unify the "we" that fights by reference to a knowledge that is ultimately veridical? How to ensure that the interdependences across different reasons for resistance remain discernable so that "we" remain woven and entangled?

From this point of view, the steadfast rejection of GMOs in Europe is an exemplary case of success. If the GMOs coming out of laboratories had continued to be defined in terms of a forward-looking innovation attesting to the capacity of humanity to master nature, we know all too well what would have happened when the consequences of their agricultural use had become undeniable. Efforts would have been made to sustain the compatibility between GMOs in the field and GMOs defined authoritatively in abstract terms under experimental conditions, even though the consequences of GMOs in the field reveal what GMOs in labs ignored: the patenting process; the increased usage of pesticides with their impact on biodiversity and health; and the proliferation of increasingly resistant "pests" (weeds, insects, fungi, etc.). The economic, ecological, or agricultural consequences would have been noted, accepted, and lamented or dismissed, but in the mode positing them as secondary effects of a rational and beneficial innovation. Even in regions of the world where GMOs have been imposed, this scenario was thwarted, and opponents have proved able to interlace their reasons for resisting: welding imagination and common sense. Each reason in isolation—from the combat against capitalist grab, through the struggle against patenting, to the defense of nature or of health—would have remained too weak, because identifiable with an ideological refusal of progress. But imagination, becoming sensitive to the reasons of others, allowed opponents to make sense in common with respect to what GMOs would mean in the field, and in a mode that caused the experts to stammer. Once activated, imagination is contagious. Today, in the wake of the rejection of GMOs, the question of what will make for sustainable agriculture continues to push us to think, to imagine, and to fight.

Problematizing Abstraction

The question of common sense has changed. Welding imagination to common sense is no longer first and foremost a task for philosophy. Today it is what activists strive for: making sense in common. Instead of coming to an agreement, making sense in common is about knowing together that the reasons for resisting, as different as they may be, need each other. Only together did activists give meaning to the agricultural innovation called GMOs. Instead of a question for specialized knowledge alone, agriculture is a question with many stakes. Ever since the Neolithic era, agriculture has entwined human practices with nonhuman forms of life, earth, and climate in a mode of irreducible interdependence, incessantly rearticulated.

To return to the image of philosophy faced with the brooding of common sense, it is imperative for philosophers to abandon the image of common sense as an attribute, something in which each human would have or should have their share, something that each human entreats us not to hold in disdain. We must imagine that the descendants of Athenian citizens today are men and women who know that no unanimous truth will be established or restored beyond their entreaty. What makes them brood is the sense that what is common is irreducibly problematic, the sense of compositions to be ongoingly revisited and always situated by what compels them to think here and now, and not in general. They feel compelled to think not as defenders of truth, but as participants in an adventure with neither destination nor heroic definition.

That said, we must also keep in mind that the image we took as a point of departure is itself situated. It is situated in terms of the descendants of Athenian citizens, who are inhabited by brooding, and they address a philosopher. We do not know what common sense might mean elsewhere. I will return to this question. At this juncture, however, I would like to share the terrain opened by activists, who have made it possible for us to make a wager against the dark will to avoid thinking, to avoid anything that might shake things up, because thinking, it is true, may inspire fear in us. The wager is that this dark will need not confront us with something we would have to acknowledge as the sad truth: that common sense would need to believe in the authority of those who know because it would be desperately incapable of distinguishing between knowledge and opinion.

This is why I wish to stress the value of experiences and knowledges that we can share with many others. When we are not in the grip of a theory, we are each and every one of us capable of juggling contextual resources, multiple practices, and semantics, according to the demands of the situation, and we are not overcome with confusion by the idea that these situations may be of importance to others in a different mode. On the contrary, that idea captures our interest, enters into our brooding, and activates our imagination. We hunger for novels that make us attentive witnesses to the passions, doubts, dreams, and fears of their protagonists. We are drawn to books about history, ethnology, and animal ethology that explore the manner in which human and nonhuman others relate to, or are related to, their world. Our imagination is indebted to fiction, which teaches us that a truth may always conceal another, and yet none of them is "merely relative."

Such an imagination is what those who have fallen into hatred have obliterated, which is why we can say they have lost common sense. This catastrophe has nothing to do with what is often held against so-called common people when they attest to what matters to them: they lack distance; they do not see the "big picture." Only the big picture would allow access to a so-called impartial conception, which would lead to condemnation of attachments whose partiality blinds them. Indeed, we are relentlessly exhorted to learn how to extract—that is, abstract—ourselves from what we hold on to, and holds on to us. This capacity to render negligible what our abstractions are abstracted from is the yardstick by which we are measured. It is the meaning of apparatuses that separate us from what we know in order to inculcate us with docility, with respect for rules defining what we have the right to know.

It is here that Whitehead may help us, because, in his vocabulary, the capacity to abstract is not the mark of a particular sort of privilege. For Whitehead, perception is in itself a triumph of abstraction; it is selective and partial, orientated by the needs of action. But, contrary to abstract thought, it does not cling to its abstractions: "The first principle of epistemology should be that the changeable, shifting aspects of our relations to nature are the primary topics for conscious observation. This is only common sense; for something can be done about them. The organic permanences survive by their own momentum: our hearts beat, our lungs absorb air, our blood circulates, our stomachs digest. It requires advanced thought to fix attention on such fundamental operations."[14] This holds true for all those animals Whitehead defines as superior, capable of observing, paying attention, interpreting. Which is to say: it holds true before language gets involved. And language became a support for abstract thought within the world of letters, dominated by the ideal of the dictionary, only when it became a matter of learning that words ought to have significance independently of the changeable, shifting relations to the reality in which they participate (with the exception of poets, who are granted latitude).

The capacity for abstraction, then, is not the privilege of thought, which allows for its authentication. Whitehead resists the judgment whose premises date back to Socrates's attempt to convince the inhabitants of Athens (without great success) that they were incapable of extracting the meaning of terms such as "courage," "virtue," or "justice" from the concrete situations that illustrate them. In other words, they were incapable of abstract thought. The dictum inscribed, it is said, at the entrance of the academy founded by Plato, the dictum to "let no one ignorant of geometry enter here," has today become a pedagogical imperative. "No one should finish primary school without severing the connection between a slice of pie (cut into equal parts) and a fraction, and the fraction should not be confused with a fractional number." When the teacher does not give students a sense of the adventure in which mathematicians are engaged, the ordeal undergone at school by children who must meet the strange demands of abstract thought is liable to produce what Stella Baruk calls "automaths."[15] Common sense is then torpedoed, blown away. Students may coolly come up with a solution for a problem in which five and a half sheep (fractional number!) are led to slaughter, "since it's all math."

Whitehead is not interested in abstract thought, but neither does he attribute concrete thought to children. Thought without abstraction does not exist for him. Thought presupposes abstraction. Of concern to him, then, are our *modes* of abstraction. In every era, writes Whitehead, a crucial task of philosophy is to *cultivate vigilance* toward the modes of abstraction that equip the thought of that era.[16]

For we never abstract in general, always in a determined mode. Thus, in the case of GMOs, their defenders see only the gain in productivity tied to genetic modification, and it is this abstraction that permits them to disqualify all opposition in one fell swoop: they have millions of mouths to feed before them. Still, it cannot be said that their opponents have access to a GMO that, for its part, would be concrete. If these opponents were able to make sense in common, it would be because their divergent modes of abstraction neither engaged in competition nor fought to establish the supremacy of one stance over another, but let each stance bring to bear a particular perspective. If modes of abstraction deserve vigilance, it is because, in every era, some of them lay claim to supremacy and relegate others to insignificance. Simply put, they lay claim to what Whitehead calls the fallacy of misplaced concreteness. This is what happened when laboratory biologists claimed to possess the scientific truth of GMOs by defining them in terms of the genetic modification that characterizes them.

The task of philosophy is not to criticize our modes of abstraction or specialized forms of knowledge mobilizing them as such. Nor is it the vocation of philosophy to oppose them with concrete knowledge. Philosophy must cultivate vigilance toward those modes of abstraction in each era that lay claim to predatory power and relegate what they omit to insignificance. To make philosophy capable of this duty, Whitehead asks that it *eliminate nothing* of what we experience, which is also to say it must never ratify the legitimacy of an omission. That does not mean defending the concrete, but rather making feel (*faire sentir*), heightening or intensifying dimensions of experience that silently insist, omitted by a mode of abstraction. In other words, Whitehead's philosophy is not militant (return to experience!), but activist, in the sense that it activates what our modes of abstraction can so easily silence.

This is why, according to Whitehead, argumentation and proof do not suit philosophy, because, whether they be scientific, juridical, or other, both rely on a particular mode of abstraction and owe their power to all that this mode of abstraction gives them the right to omit. Yet philosophy can never give itself such a right. It cannot take pride in the high fact of explanation through eliminating what its explanation needs to neglect. In mathematics, however, proof is nothing but a path for a type of comprehension bearing its own self-evidence: "The attempt of any philosophic discourse should be to produce self-evidence. Of course it is impossible to achieve any such aim. But, nonetheless, all inference in philosophy is a sign of that imperfection which clings to all human endeavor. The aim of philosophy is sheer disclosure."[17]

"Disclosure" does not correspond to "enlightenment," with its connotation of unveiling a truth that had till then been dissimulated (or repressed). Something becomes manifest, but it is not a matter of manifestation in the phenomenological sense (in the manner of "the appearing of the entity itself in its manifest being"). Manifestation is not in opposition to abstract capture. It establishes itself for itself, purely and simply, and another mode of capture is established with it that gives importance to what we habitually omitted and asks us to pause on what seemed anecdotal. In other words, the openly affirmed aim of philosophy should be to activate dimensions of experience that our modes of perceptual and linguistic abstraction omit, and it is not for nothing that, in *Modes of Thought*, Whitehead dares to speak of the kinship between philosophy and poetry. In doing so, he does not claim that philosophical statements are in need of poetic quality. "Philosophy is the endeavor to find a conventional phraseology for the vivid suggestiveness of the poet,"[18] to craft statements "in prose," and even "prosaic" ones, that are liable to impart to what is usually omitted the power of making consciousness "flicker": "Our enjoyment of actuality is a realization of worth, good or bad. It is a value experience. Its basic expression is 'Have a care, here is something that matters!' Yes—that is the best phrase—the primary glimmering of consciousness reveals something that matters."[19]

The task Whitehead proposes for philosophy—vigilance toward our modes of abstraction—does not correspond to the indictment of a consciousness that would remain desperately imprisoned by ultimately inadequate modes of abstraction without the intervention of philosophy. To the contrary, it is when something suddenly demands us to pay attention (something that we had not felt important until then) that we feel ourselves conscious. For Whitehead, consciousness is not a stable attribute; it flickers, embarking on a permanent adventure in modes of abstraction distributing what matters and what may be omitted under such and such a circumstance. And it may be the taste for this adventure that the imagination, set in motion by fiction, translates. As for those who wish to enclose us within the finite framework of what we may legitimately know, whatever they may say, we know very well that a truth can always conceal another within it.

While the task of philosophy is to weld, welding does not necessarily happen between two disparate realities, common sense prisoner to its routines versus free imagination. In fact, after proposing the welding of imagination and common sense, Whitehead defines the aim of this welding: it should produce "a restraint upon specialists" as well as an "enlargement of their imaginations."[20] Restraint should not evoke the image of a foot slamming on the brakes, or the image of common sense relentlessly policing an excess of speed. It is more a matter of making evident the unrestrained advance of a form of knowledge demanding protection from anything that might slow it, harnessing the imagination of those who devote themselves to its advance. And here Whitehead takes very particular aim at those modern specialists whom he dubs "professionals." Recall the biologists who created GMOs. They knew very well that things might happen in the fields that they had not observed in the laboratory, but that did not activate their imagination. They insisted that everything will be worked out, must be worked out, because these sorts of complication are not worthy of stopping progress. "Each profession makes progress, but it is progress in its own groove. . . . The groove prevents straying across country, and the abstraction abstracts from something to which no further attention is paid. . . . The remainder of life is treated superficially, with the imperfect categories of thought derived from one profession."[21]

The specialized modes of abstraction of our modern professionals give a great deal of importance to one of the order-words of modernity, "progress." While these modes of abstraction (not surprisingly) clash over the meaning of progress, they agree on one thing: the need to break with common sense. They function in a manner that can be qualified as predatory, for they lay claim to the power to disqualify and silence others. Yet they cannot be defined as inherently predatory, for it is the professionals who confer this predatory power on them, wherever these professionals are recognized as servants of progress, of course. A specialized mode of abstraction can never be said to be *inherently* guilty. Still, in the manner of Donna Haraway, we may say that they are *not innocent.* Vigilance is required vis-à-vis all claims to innocence and vis-à-vis all arguments presupposing that it is legitimate to eliminate or overlook whatever a mode of abstraction does not deem to be of importance.

If no mode of abstraction is innocent, it is not justified to consider them all suspect, all guilty, as if becoming professional were the inevitable destiny of the specialist. Whitehead often advises us not to exaggerate. He thus engages in a form of humor, a humor we might call prosaic. Humor does not disqualify its targets; its triumph lies in spurring the reader to laugh not only at herself but also at the manner in which she has become accustomed to (re)presenting herself. Humor is aimed at professionals who, although cognizant that their modes of abstraction dominate them, continue to think of themselves as paying the price demanded by their so-called vocation. This view of a vocation is rather exaggerated: in older articulations, the vocation does not imply blind engagement, but quite the contrary. We need not fear that, by asking specialists to acknowledge a world beyond their groove, we are destroying vocations. Whitehead stresses that the construction of professionals, or fixed persons with fixed duties, is a modern invention, no older than the twentieth century. Obviously, fixed persons with fixed duties as such are nothing new. Antiquity had its professionals in scribes, officials, and astronomers, all figures of scrupulous and short-sighted precision. What is new, Whitehead notes, is the coupling of profession and progress. This coupling involves the invention of institutions in which entrepreneurial yet fragile imaginations are cultivated, which demand protection against questions that *must not* concern them.

Professionalization thus raises questions about the milieu it requires—that is, its associated institutional milieu—and about this other milieu that must let itself be kept at a distance, a milieu in which respect must be imposed for a knowledge whose authority constructs a void around it. For a mathematician like Whitehead, mathematics does not need to be authoritative, because the values it brings into existence do not need to deny anything at all; values are established through their very self-evidence. It is the educational milieu in which mathematics is inculcated that tends to produce "automaths" who are blindly obedient to definitions. The same is true of the facts mobilized for argumentation and proof. When facts are required for someone to be heard, their functioning can turn predatory: "where are your facts?" In French, the word fact, *fait*, can take on another intonation. When workers or artists stand back and consider their accomplishment, they say: "there, it's done"; *c'est fait*. Here, the fact is something done, and at stake is the difference between something done well or done poorly. This understanding of fact, however, runs into the ironic request for definitions. What is meant by "done well"? It is then readily accused of entertaining a reactionary and/or Romantic attachment to allegedly true values, or else of indulging relativism and subjectivism. Our milieus are toxic. Generally speaking, the predatory functioning of specialized modes of abstraction is a matter of milieu, and in particular, the milieu associated with the invention of the modern professional.

The philosophy of Whitehead is thus situated by a preoccupation with that which characterizes the milieus proper to the modern world, those that made philosophy itself into a professional activity (especially by producing pared-down statements haunted by the ideal of the dictionary). Whitehead's philosophy is not rooted in the ground of a truth to be deployed. Neither does it aim for an ideal transcending situatedness as the very vocation of thought. His philosophy disavows the critical weapon allowing for the identification and denunciation of what corrupts thought. This philosophy dovetails with the radical meaning of the doctrine of evolution, which, as Whitehead puts it, knows neither foundation nor ideal, and which does not understand finitude in terms of limitation, because limitation always implies a reference to what it deems inaccessible, a reason that would demonstrate that what is had to be thus. To accept the doctrine of evolution is to affirm that what exists could have existed otherwise, but without falling into the unhappy irony of relativism, and without repeating the melancholic ideal of truth without attachment, ground, or history whose impossibility he announces. The naturalist who contemplates the marvels of a spider weaving its web may be intrigued, but he does not cultivate the least nostalgia for a creator God who wanted this spider to be what it is. Whitehead makes the wager that common sense may be intrigued by specialized knowledges without needing to lend them an authority to which it should bow. And of course, to avoid being a pious vow, this wager calls for a change of milieu.

Civilizing Modernity?

Obviously, Whitehead did not foresee what we live today. I chose to situate him in the ruins of Athens, an emblematic site for the origin stories of philosophy, but without any ambition to discuss this origin or any desire to give Whitehead's addendum to this philosophical scene the power to offer answers to today's questions. That would make him into a clairvoyant or prophet. In fact, we can say that reading Whitehead demands us at once to situate him and to situate us. Whitehead belongs to his era, with anxieties that are still our own, and with what may appear to us as blind spots.

Indeed, reading Whitehead, we may at times be tempted to adopt the stupid stance of superiority of the present over the past: we know better; we would speak otherwise. In this essay, I have chosen to impart greater density to what I share with him through more contemporary themes. As for blind spots, we would do well to remember that the issues that we are no longer supposed to ignore today are not necessarily ones we now prove capable of thinking in a coherent manner. Instead, what haunt us are debts that have till now remained unpaid. Perhaps, as Étienne Souriau would say, we have not (or not yet?) paid the price for our passage. Perhaps the facility with which we are tempted to burn today what yesterday we adored is a translation of the temptation to consider critique as an accomplishment in itself, even when the problem has not been effectively or positively deployed.

We do not yet know, then, whether the possibility inspiring Whitehead to think—civilizing modernity—is feasible, or whether we are living through the collapse of this civilization. We do not know whether the stories to be told about us in the future will build on the continuities of a living heritage, or whether they will speak of us in the mode of "them," strange creatures who bequeathed a world in ruins.

To be sure, reading Whitehead today pushes us to accentuate the question mark hovering over the very possibility of civilizing modernity, but we do not need to teach Whitehead about the speculative nature of his wager; far from it. In fact, what pushed Whitehead the mathematician into philosophy was the conviction that modern civilization was in decline. This conviction might be called prephilosophical, forged by an academic who asked himself how and whether this civilization could ever recover from the terrible climate in which industrial society was born, an academic who knew full well that most of his lettered colleagues, prisoners of professionalized modes of abstraction, participated in this climate and responded to the question of working class misery with the answer Cain gave to God: "Am I my brother's keeper?"[22] As we have seen, the positive task Whitehead assigns to philosophy, the task of welding common sense to imagination, is aimed at these specialists whose unrestrained confidence in progress needs to be checked, and whose imagination needs to be enlarged.

It might be said that a brooding common sense has turned Whitehead into a philosopher, with the task of elucidating what specialized forms of judgment tear asunder, and of giving coherent meaning to it. The figure of philosopher, then, is not that of a revolutionary. Even though revolutionaries are right to denounce the relationship between the professionalization of knowledge and social oppression, to style philosophy as revolutionary would once again give a particular mode of abstraction the power to define the truth of a situation.

Common sense of the sort Whitehead highlights does not ask, "who is right?" Common sense asks for coherence where contradiction reigns. But what coherence can be wrought from these specialized modes of abstraction in dispute among themselves? Whitehead, the mathematician-philosopher, answers that coherence will not define any form of "correct thinking." In response to Socrates's question "what is . . . ?," it won't do to answer with the idea that, if each mode of abstraction remained within its proper field of application, they would work together, in accordance with a peaceful and courteous division of labor. Disagreements are not to be written off as losses, distortions, or defects. For a mathematician, coherence is not something to be retrieved. Coherence is to be created.

Just as welding is a metallurgical operation, Whitehead will take the creation of coherence as a technical operation. Such an operation has for its aim to endow philosophy with the capacity to activate the importance of what "we know," even if specialists refuse our right to let our knowledge matter. If we are to civilize the modes of abstraction that modern thinkers have dedicated to battling each other, their only common goal being to disqualify common sense, we must reevaluate what these modes of abstraction call for, in such a mode that they lose any relation to right or legitimacy. Such reevaluation will allow us to approach them in the manner of adventurous modes of capture that are always at risk of neglecting something that, here, in this case, might matter. A mathematician would ask no less, and Whitehead goes to great lengths to accomplish as much in his metaphysical work *Process and Reality*.

Civilizing modernity, then, means getting specialists to learn to situate themselves, to use Haraway's turn of phrase, or, to evoke Deleuze, to honor the truth of the relative (in contrast to the relativity of the truth), the truth of knowledges that know how to present themselves as relative to the questions they prove able to pose effectively. Both approaches urge specialists to make connections actively between what they know and what their knowledge must omit in order to be produced. That seems like nothing, you may say, but let's recall the GMO affair. Specialists speaking about GMOs knew quite well that they had omitted any questions that could not be addressed in their laboratories, but it took activist struggles to put an end to the presentation of GMOs as the key to such a decisive advance in agriculture that other questions were more or less negligible, to be dealt with later. In this instance, civilizing modernity would have meant reuniting all the protagonists concerned to make sense in common about what was claimed to be coming out of the laboratory. Above all, it would have meant that, when presenting themselves or their specialized knowledge, everyone would have avoided claiming for themselves terms such as "universal," "objective," and "rational." In one way or another, such terms all consign those to whom one is speaking to the dark realms of anecdotal knowledge, particularistic habits, subjective attachments, irrational beliefs, and impassioned affect. Civilizing modernity is quite a worthy ambition, deserving of its question mark.

But those activists had not read Whitehead, you might say. It was their struggles that welded an enlivened imagination to the question of a sustainable agriculture, a question now shared by many. As we shall see later, Whitehead's metaphysics may be suited to the "making sense in common" generated by activists, which extends the notion of common sense not only far beyond the streets of Athens, but also far beyond the image of each human entitled to "his" opinion. Driven by the question of coherence, the singularity of Whitehead's metaphysical system forced him onto a literally experimental path where he made and unmade his concepts to the point where not one of them could hold independently of the others.[23] In doing so, he rid himself of the ways of posing problems and the economies of thought that his era authorized. This does not mean that he attained the universal above and beyond the limitations of his era. Whitehead's system is situated by the very thing it aims to create, coherence among our modern modes of abstraction. Thus, we might say that he is no longer situated by an era, but by the question of civilization to which his era belonged, and ours as well.

This civilization is in question, and not *civilization*. Whitehead was always attentive to the plurality of civilizations. He likes to contrast modern civilization, born in Europe, with that of the Greeks, the Egyptians, or the Semites. Yet he never develops the idea of a progress of *civilization* across civilizations that would only constitute stages of it. In *Modes of Thought*, the notion of civilization is presented as a generality, and in *Adventures of Ideas* as "difficult to define,"[24] which is true of all generality in Whitehead's sense. But he never wavers on one point: there is no model civilization. To copy the Greeks, for instance, would have no meaning, because they themselves were not copyists: "They were speculative, adventurous, eager for novelty."[25]

What Whitehead means by civilization might be called a very special milieu of culture for the adventures proper to human life, within which traditions are not only cultivated but also put to the test and set out on adventures. Each civilization, writes Whitehead, "deposits its message as to the secret character of the nature of things,"[26] as to what it means to live, to act, to feel, and as to what meaning to give to order, as well as disorder, conflict, and frustration. Thus, each civilization proposes to human experience a manner of understanding itself that includes disagreements, imparting tension to civilization itself, as well as conflicts tearing it apart. Civilization in this sense is keeping with the stakes that constitute Whiteheadian common sense: common sense is not some fund of knowledge common to humanity, but a situated capacity to participate in the adventure of civilization to which it belongs.

It could be said that, for Whitehead, a civilization is an adventure that is thinking itself through those who are concerned with it. It is in this sense that he could write, "civilizations can be understood only by the civilized."[27] There is, then, a radical historicism in Whitehead's thought, but it is an historicism bearing on what matters for this civilization, what makes those who belong to it think, and what aspects of existence are liable to make them brood. Their civilization, its future or decline, *concerns* the civilized. For them it is a question that makes them think, hesitate, judge, hope, or slaughter each other. And as a philosopher, Whitehead does not claim to transcend the civilization to which he belongs. The challenge he assigns to philosophy, welding imagination to common sense, would undoubtedly have been quite difficult to explain to an ancient Greek. Yet, insofar as it seems to us at the very least debatable, we are contemporaries of Whitehead from this point of view.

It is important to emphasize that Whitehead's historicism does not at all imply that we should understand his thought as deriving from his historical situation. Rather, his historical situation forced him to think and to create concepts that activated his problematization of it. The conceptual adventure of Whitehead is, in this sense, like a message in a bottle thrown into the ocean, or like an arrow shot into a future he will never know: this is what happened to us, this is how the civilization to which I belong has allowed me to understand its decline.

We may feel today that the modern world can no longer be characterized in terms of decline. Instead, we are faced with the question that historians of the present call "the great acceleration," in reference to the accelerated intensification of the impact of "development" on our environments and ourselves. But this fact is not, as such, liable to call into question what made Whitehead think. Capitalism, which drives what we call development, can certainly explain the manner in which this kind of development has been unable to think its consequences. But capitalism does not explain the fact that, until very recently, so many knowledgeable people have identified development with progress, even if it meant waiting for the end of capitalism for this progress finally to benefit the whole human race. Today we can but note that what such development has provoked in a sense verifies Whitehead's diagnosis against modern civilization: the art of imagining has been monopolized and harnessed by specialists become professionals, deaf to uneasiness and protest. Those who feel uneasy and protest have been accused of seeking to impede the inevitable march of progress—Pascal Lamy's statement still rings in my head: "You can't stop the clocks." Whitehead wrote that all societies (which is also to say, as we shall see, all that endures as it endures) rely on the patience of their environment as to the manner in which they affect it. The earth has lost patience, and the ticking of clocks has become ominous.

Walking this path with Whitehead, I have chosen to follow him wherever I could do so, in the mode of "I concur and would say even more." Where I feel my era has not only permitted me to say more but also asked me to speak otherwise, I will take up the baton from him. Taking up his baton does not concern his conceptual system, for as we will see, it functions through its own constraints in the manner of a machine to make think. Modifying it would mean reconstructing it, an undertaking for which I do not have the least ambition, since I owe to Whitehead my capacity to think with those whom I feel to be my contemporaries. It is Whitehead who let me dare to venture with them into this zone of indiscernibility where we no longer know whether the language we speak is still that of the modern era. I will nonetheless focus on the texts, principally *Modes of Thought*, in which Whitehead, heir to his own conceptual adventure, addressed his own contemporaries, sought to share with them the taste of a thought that does not submit to what we have the right to know. Take these two passages:

> In mankind, the dominant dependence on bodily functioning seems still there. And yet the life of a human being receives its worth, its importance, from the way in which unrealized ideals shape its purposes and tinge its actions. The distinction between men and animals is in one sense only a difference in degree. But the extent of the degree makes all the difference. The Rubicon has been crossed.[28]

> The central organism which is the soul of a man is mainly concerned with the trivialities of human existence. It does not easily meditate upon the activities of fundamental bodily functions. Instead of fixing attention on the bodily digestion of vegetable food, it catches the gleam of the sunlight as it falls on the foliage. It nurtures poetry. Men are the children of the Universe, with foolish enterprises and irrational hopes. A tree sticks to its business of mere survival; and so does an oyster with some minor divergences. In this way, the life aim at survival is modified into the human aim at survival for diversified worthwhile experience.[29]

In his era, Whitehead must have scandalized those thinkers who defined the human as a thinking being, as *sapiens*. For him, if the Rubicon has been crossed, it is because an "outrageous novelty is introduced, sometimes beatified, sometimes damned, and sometimes literally patented or protected by copyright,"[30] which attests that what might be and yet is not has the power to insist and that the importance of unrealized alternatives has been introduced into the world. Whitehead never ceases to remind us that, without this sense of possibility transfiguring the "given" or "datum," there would be no morality, no religion, no technique; nor would there be science (a fact we too often forget). Nor, for that matter, philosophy. There would be no common sense to brood over aspects of existence. After all, if common sense broods, it is because it seeks more than what is given. To be sure, characterizing humans in terms of their foolish enterprises and irrational hopes presents an interesting departure from discourses on rationality proper to the one who is christened *sapiens*, but it is equally clear that Whitehead was not bothered by the questions haunting us today. The currently pressing question about human exceptionalism that arises in our relations to animals does not stop with them. It may also concern trees, which do a great deal more than merely survive. The very image of the crossing of the Rubicon once brought a furrow to Donna Haraway's brow. For all Whitehead's liberality in granting empirical knowledge and emotions to so-called superior animals, Haraway heard in it echoes of a separation without return, of the founding of Empire, and of History, whose only protagonists were now human beings. As for the foolish enterprises and sensibility for unrealized alternatives of these "human beings," we cannot and must not forget today that they demand much more than the vigilance of philosophy. Can we, even in a poetic fashion, characterize the human as such in terms that belong first and foremost to our civilization? Let us recall the ominous juridical thesis of *terra nullius* that defined certain lands as belonging to no one because they were inhabited by "lazy" people, strangers to the spirit of enterprise, who had not put their lands to work. It designated them as peoples whose "traditions" should be destroyed, since they stifled the insistence of unrealized alternatives that gives its value to human life. Can we say that it was the human who crossed the Rubicon, if that means separating *nature*, which *is*, from *culture*, which nurtures what *might be*? Or was it the colonizer, in charge of the "foolish" mission of civilizing humanity, while animated by the possibility of taking possession of worlds to exploit?

Still, we should not then conclude that we now know what Whitehead's belonging to his era allowed him to ignore. Our contemporary questions that his era did not pose are less what makes us think today than they are what puts us to the test. Denunciation of the past cannot serve as thought, nor can guilt serve as principle.

And, as always with Whitehead, we must not move too quickly. We must hearken to the call that impels him as a philosopher to take the side of common sense against the absurdity of doctrines that take so much pride in relegating to arbitrary subjectivity not only that to which they owe their existence—faith in the possible—but also that which animates them when they preach the necessity of judging things objectively: the unrealized alternative of a humanity finally liberated from its illusions. His call is: "We are children of the universe." We have become so accustomed to thinking that the universe is indifferent to what makes us thrill that we recognize any thesis whatsoever as "undoubtedly objective" so long as it has the allure of a "truth that hurts," a truth translatable into the defeat of commonsense. It is Whitehead's metaphysical adventure that allows him to issue this call, and his metaphysics says nothing about the human. It is the universe itself, as his metaphysics conceives of it, that gives meaning to the possible, to the insistence on unrealized alternatives, to the link between existence and value, and all of it in a mode that may concern the oyster and tree as much as the human. "We have no right to deface the value of experience which is the very essence of the universe."[31]

"We are children of the universe" thus brings us into contact with a duty, an engagement against the absurdity in which modern philosophy has participated in the foolish enterprise of making modern entrepreneurs the very prototype for the human coming at last to the "age of reason." Whitehead himself did not go any farther. He did not inquire into those different manners of respecting the value of experience that belong to other children of the universe. But today as yesterday, his metaphysics can help us loosen the grip of the unavoidable dilemmas that are strangling us.

{ 2 }

In the Grip of Bifurcation

Whitehead has no intention of bringing reason to bear on our irrational hopes and foolish enterprises. To civilize is neither to domesticate nor to placate. Whitehead does not propose to define what we know to build consensus. Giving common sense the power to resist does not mean giving precedence to concrete knowledge in opposition to the abstraction of specialized forms of knowledge. What allows specialized forms of knowledge to work is not abstraction per se, but the authority claimed by certain modes of abstraction, which has the capacity to separate us from what we know.

The Bifurcation of Nature

Whitehead the mathematician turned to philosophy in a bid to bring coherence to what for him was the flashpoint of modern thought, what he calls the "bifurcation" of nature. The bifurcation of nature, separating nature into two distinct registers, generates modes of abstraction leading to what he considered pure and simple absurdity, the mother, as it were, of all the battles waged against common sense. On the one hand, there is an objective nature, ruled by causality. It causes our perceptual experience in particular. On the other hand, there is nature as we perceive it, rich in sounds, colors, and odors, as well as values, emotions, fear, and wrath. It is mere appearance, for which the human mind alone would be responsible. Those who dare admire a sunset, then, are admiring what they themselves bring into existence. "The poets are entirely mistaken. They should address their lyrics to themselves and should turn them into odes of self-congratulation on the excellency of the human mind."[1]

Fighting against such absurdity does not mean promoting truth. To stick with this bifurcation of nature regardless of the absurdities it generates offers a powerful example of foolish enterprise and irrational hope. Yet the example itself does not point us in the direction of wisdom and reason. In this instance, Whitehead wants to make us aware of a pure and simple abuse of power to be rejected as such. Thus, refusing any negotiation with the bifurcation of nature, he writes: "Everything perceived is in nature. We may not pick and choose. For us the red glow of the sunset should be as much part of nature as are the molecules and electric waves by which men of science would explain the phenomenon."[2]

Let us not mistake his point. Whitehead is not announcing a project to reconcile what is irreconcilable. His protest concerns inconsistency in the mathematical sense, an arbitrary disjunction between principles of intelligibility. Each principle is given as self-sufficient, as if each could be defined independently of the other, when in fact each is defined in opposition to the other. Objective nature is defined as independent of mind, while also being an object for consciousness, capable of prevailing over appearances. Apparent nature, for its part, would be quite capable of laying claim to all that we know, making the human mind alone responsible for the knowing of molecules and electromagnetic waves. Whitehead's turn of phrase, to "be as much part," affirms the necessity of creating a coherent conception of nature, not trying to overcome an opposition that derides any possible conciliation.

In his first genuinely philosophical work, *The Concept of Nature*, then, Whitehead does not attempt to conceive a nature that would be at once the one whose beauty the poet celebrates and the one that scientists explain objectively. He creates a concept of nature: that of which we are aware in perception. In Whitehead's articulation, perception is not a matter of what is perceived by the human mind. Perception concerns "that of which" there is experience. Nature thus defined is neither in itself nor for us. Nature is conceived in terms of the "grasp" it affords for the variety of perceptual experiences, be it the experience of poetry, the scientist's experience, or the rabbit about to leap. The concept of nature is designed to forestall any contradiction between an objective nature that would be responsible for our perceptions and an apparent or subjective nature that would refer to our own responsibility. It highlights diversity of kinds of grasp, modes of abstraction to which nature is susceptible.

Throughout his career, Whitehead's principal interlocutors were Newton, Hume, and Kant, three thinkers who, in his view, had contributed immensely to instituting the bifurcation of nature, constituting it as the insurmountable horizon of thought.

I will largely dwell on Newton here, because it is Newton, and the line of mathematical physicists till Einstein who followed in his footsteps, who expanded the authority of the bifurcation of nature well beyond the bounds of the philosophical distinction between primary and secondary qualities already promoted by the atomists. The establishment of Newton as a figure of almost legendary proportions, independently of his actual ideas, insights, and aims, can be associated with the triumphant discovery of an objective nature, ruled by universal laws that could be formulated through simple observation and calculation. Pierre-Simon Laplace captured this triumph in his famous response to Napoleon: "There will never be a second Newton because there was only one world to discover." For Whitehead, Newton's triumph lies in the notion of self-sufficient fact. Newton could stick to the facts: the observation of successive positions over time of a certain planet proved sufficient for determining its orbit through increasingly complex calculations. Additional observations proved sufficient to show that this orbit obeyed the same law that governed the movement of all other planets, as well as the trajectory of the mysterious comet whose periodic return to our skies had been an object of observation for centuries.

For physicists, the idea that nature is "knowable as it is in itself" does not pose a philosophical problem. Because nature is typically associated with universal laws, no one raises an eyebrow when someone declares that, if extraterrestrials somewhere in the universe were to develop an active interest in the functioning of reality and were to learn how to observe and calculate, they would inevitably formulate the same laws. To call on extraterrestrials is to evoke beings whose modes of thought and feeling, culture and ideas, are utterly unknown to us. We have no idea what apparent nature would be for them. Facts, however, are sufficient in themselves, and the rest may be bracketed: it suffices that these extraterrestrials be able to observe and capable of mathematical reason for them to reach an agreement about the physical reality they study.[3]

Of course, other laws have replaced Newton's for characterizing "the one world to discover." Max Planck, who introduced the extraterrestrial argument, enthroned the conservation of energy. Others today propose the quantum theory of fields. Needless to say, the general relativity of Einstein still enjoys pride of place, resituating as it does Newtonian physics in terms of an approximation valid only for slow and heavy bodies.

Whitehead's historical moment saw the refutation of the Newtonian theoretical edifice, till then thought to be unshakeable. Whitehead compared this episode to a Greek tragedy, whose essence, he wrote, "resides in the solemnity of the remorseless working of things."[4] But Einstein's new universe did not convince Whitehead with its absorption of universal gravitation into the geometry of space-time. As early as 1922, Whitehead proposed an alternative to Einstein's general relativity, whose conceptual structure failed to satisfy him.

Other theoretical alternatives were proposed at the time. Some specialists maintain, however, that the one proposed by Whitehead still holds. That his theory has not captured the interest of physicists is not all that surprising, for in contrast with Einstein, Whitehead's alternative calls into question the triumphant attainment of universal laws of physics. In effect, Whitehead's theory quietly proposes that physics become a science among others. He proposes that physics, like other sciences, is dealing with entities that affect their milieu and are, in turn, affected by it. According to this approach, it belongs to each body that physicists associate with a mass to define its own spatiotemporal stratification, its own discrimination between space and time. Gravitation, then, no longer communicates with a law to which the movement of a body is subjected. Gravitation is the manifestation of relations between stratifications defined by those different bodies, each on its own account. In other words, Whitehead's theory does not merely reject the absorption of Newton's gravitational physics into Einsteinian space-time. It calls into question the universality of gravitational physics as such.

Whitehead did not seem surprised that his alternative formulation of general relativity won the appreciation of only a handful of experts. In 1922, he took the path that would make a philosopher of him. The important thing for him was to have shown that general relativity, as innovative and pertinent as it may be, did not require the universe to be "reduced to static futility—devoid of life and motion," to a changeless order "conceived as the final perfection."[5]

Yet Newton's laws have been applied well beyond the domain that first attested to this immutable order. Laplace invented his famous demon in reference to the perfect knowledge that the Newtonian order makes possible. In cases in which we have to resort to probabilities, in the so very numerous cases when it is not possible to realize the deterministic forecasting associated with the Newtonian order, the demon provides a justification. With his demon, Laplace evoked a pure intelligence capable, in the manner of an astronomer, of contemplating any instantaneous state of the (Newtonian) universe, and thus of deriving the past and the future of this universe from the definition of that state. Probabilities do not belong to Newtonian physical reality, but thanks to Laplace's demon, they do not call it into question: it is only because our human means of observation and calculation are approximate that the course of things appears indeterminate to us. His demon enables a situation where "one objective nature" rules beyond "appearances," whatever the challenging variety of the so-called appearances. It is a gesture that has been repeated any number of times. Today we hear neurophysiologists affirm: if we could define the instantaneous state of someone's central nervous system in terms of interactions among all its components, we would be able to deduce from this definition his or her experience in that instant; we would at last relegate all ideas about liberty and spontaneity to the outdated past. Naturally, neurophysiologists have not enjoyed the same success as contemporary physics, but they can recycle the ideal of perfection associated with Laplace's demon: "if only we could . . ."

Exorcising this idea of perfection, then, would involve dramatizing the fact that Laplace's demon, supposed to be able to foresee all, not only would reduce what matters to us to static futility but also would not provide answers to any of our questions. In effect, from the so-called omniscient point of view, nothing poses any questions. Our questions attest to our imperfect knowledge. This is why Whitehead insisted that the ideal of perfection results in pure tautology, a succession of states defined in terms of their very equivalence.[6] The demon seems to have escaped appearances, to have exited Plato's well-known cave, but in the physicist's version, once outside, he distinguishes nothing. In a daze, as if blinded by the sun, he is incapable of wondering about anything whatsoever. Thus, Whitehead writes with respect to the physics of his era, it can only repeat, "that is so,"[7] in the form of a "mystic chant."[8]

The bifurcation of nature would find its second thinker in the person of David Hume. Hume envisaged a perceptual field meticulously purified of anything that might afford access to a self-sufficient reality, holding together by itself. Here apparent nature has the upper hand. In the end, humans are responsible for constructing interpretations about how nature holds together. Nothing of what we call nature may claim the least intrinsic intelligibility. In particular, nature does not authorize the objective identification of causes and effects. "Pure sense perception does not provide the data for its own interpretation."[9]

Whitehead addresses Hume as if he is a fellow philosopher who has plainly exaggerated to make his case. In order to deny that reality has the power of causing anything whatsoever, Hume must engage in make-believe. When he argues, he claims that visual sensations are born in our souls by unknown causes. Elsewhere, however, when he forgets himself, Hume writes what everyone well knows: it is through the eyes that we see. And Whitehead remarks: "The causes are not a bit 'unknown,' and among them there is usually to be found the efficacy of the eyes. If Hume had stopped to investigate the alternative causes for the occurrence of visual sensations—for example, eyesight, or excessive consumption of alcohol—he might have hesitated in his profession of ignorance. If the causes be indeed unknown, it is absurd to bother about eyesight and intoxication. The reason for the existence of oculists and prohibitionists is that various causes are known."[10]

The entrance of oculists and prohibitionists onto the scene as partners in a philosophical argument may well startle the serious philosopher. It is as if Whitehead makes an alliance with common sense, or even with the Thracian servant girl who famously made fun of Thales when he fell into a well, too absorbed by the heavens to see what was directly beneath his feet. Whitehead is not particularly impressed by the grand battle carried out by Hume against attributing the cause of vision to what we see. For Whitehead, the efficacy of the eyes does not present a daunting challenge. "For example, 'I see a blue stain out there,' implies the privacy of the ego and the externality of 'out there.' There is the presupposition of 'me' and the world beyond. But consciousness is concentrated on the quality blue in that position. Nothing can be more simple or more abstract. And yet unless the physicist and physiologist are talking nonsense, there is a terrific tale of complex activity omitted in the abstraction."[11] The omission of complex activity proves all the harder to sustain because Hume must not even allow himself to perceive, for instance, blue clothing: the fact "of blue, out there" must remain sterile, not pure, but actively purified of all that invites interpretation. Sensualism is hyperintellectualism.

Is it possible to be heir to both Newton and Hume? We thus arrive at Kant, "the first philosopher who in this way combined Newton and Hume. He accepted them both, and his three Critiques were his endeavor to render intelligible this Hume-Newton situation."[12] Kant may be credited not only with establishing the bifurcation of nature as doctrine. He invented another doctrine of bifurcation, within human experience as such, between the empirical experience of values and moral law, empty yet imperative.

It is worth noting that Kant ratified the universal character of the mode of explanation associated with Newtonian physics. But this mode of explanation no longer refers to the world as it exists independently of the perceiver. It characterizes how the Kantian subject constitutes the object it perceives. In other words, the Newtonian achievement is neither speaking about the world nor explaining it. For Kant, his genuine achievement is to have disclosed and put to work categories that correspond to the constitutive principles of any perceived object. The perceived facts seem sufficient to define the object because they answer to categories of perception. As such, they are bound to provide evidence of the object in terms of those principles.

Not surprisingly, physicists welcomed this news somewhat coldly. But it was a godsend for sciences that strove to imitate physics without being able to match its achievements. It allowed them to claim:

"We have no choice; the only rational approach to what we are dealing with is to follow the example of physics." Such sciences no longer need resort to the infinite knowledge of Laplace's demon. A form of principled determinism suffices to introduce the bifurcation between the empirical, abounding with accidents and events—Life and Movement, as Whitehead often says—and its rational intelligibility. Each science has then to put into words categories that do no more than explain the principles "pre-forming" its object. But it is here that Kant introduced a supplementary bifurcation that steps in to protect morality. The danger is that the subject, if it is described rationally, will be understood in terms of what determines it. The subject will then be understood as inherently irresponsible; its most heroic or criminal decisions will be causally explainable. It will be a matter of preserving responsibility instead of what matters or has a sense of value. Kant thus institutes a transcendental subject that stands outside the stuff of causality, ensuring that an act can be imputed to the one who committed it. This imputation puts in place a pure responsibility: no excuses, no good intentions, no explanations. The bifurcation has made a clean cut. Responsibility remains unstained by empirical attachments. Indifferent to circumstances and reasons, the inflexible authority of universal moral law declares, "You should not have!"

Kant, I might add, not only combined Newton and Hume. He also established domains of territorial legitimacy and proscribed all smuggling operations. The Three Critiques ensure exclusive rights of sovereignty to the tenants of separate territories, each ruled by its own Critique. What had been the domain of Truth is now a matter of Pure Reason, while Good is now dependent on Practical Reason, and Beauty on Judgment. Here we see the grasp and the stability of the regime of bifurcation. Bifurcation provides the conditions for the guardianship of each territory, as well as for the attempts at annexation to which the territory is sporadically subjected. We are forbidden from taking into account forms of knowledge running counter to the framework thus instituted. As for common sense, it must remain outside the game. Thus, when neurophysiologists claim to naturalize the mind and the outraged guardians of human subjectivity mobilize themselves, common sense is a perplexed witness to the conflict between the representatives of the irresistible advance of science and the heroic defenders of a subject attacked by barbaric powers. Territorial peace is an armed peace. The only point of agreement between the confronting sides is that people, if they are meant to take part, are supposed to learn to distinguish legitimate voices from those of impostors. People are not to become involved with the defining of territories.

One may nonetheless argue that Newton, Hume, and Kant are far from us today. Why do I feel I can describe them as if their power remained intact?

The Trick of Evil

"Surely you believe that physical reality exists independently of what we think about it?" During the renowned science wars, critical thinkers contended with this sort of challenging question, sometimes accompanied with a proposition to throw themselves out the window on the fifth floor, or even higher. In the previous chapter, I have underscored the rather aberrant nature of this proposition, which implies that one of the achievements of modern physics had been to establish that such a fall would have disastrous consequences. Now I turn to another aspect of this provocation. The fall of a heavy body was proposed instead of, say, the precise date and time of the next solar eclipse and the regions of the world where it would be visible. It seems that the successful outcome of such a forecast does not matter. Even the most ignorant human or the first dog to come along is aware of the necessity of not confusing a door with a window. For the physicists, however, when they are at war, the predictable crash belongs to physical reality just as much as celestial mechanics do. "Physical reality" is how they refer to nature independent of the human mind, the nature that is what it is. Simply put, their challenge brings into play the bifurcation of nature.

I would like to develop an association between the thought-crushing manner in which the physicist plans to back the doubter into a corner and what Whitehead characterized as the "trick of evil," when something new claiming consideration arouses a furious and dogmatic rejection. "Insistence on birth at the wrong season is the trick of evil. In other words, the novel fact may throw back, inhibit, and delay."[13]

The same idea appears in *Religion in the Making*, but in the guise of an evil that promotes its own elimination as felt painfully at cross purposes with what feels "good." Evil is eliminated when the capacity to feel the pain is destroyed, which is also the capacity to admit into feeling what threatens the definition of what is good. Which means partial anesthesia, when we sadly accept what was previously intolerable. "It must be noted that the state of degradation to which evil leads, when accomplished, is not in itself evil, except by comparison with what might have been. A hog is not an evil beast, but when a man is degraded to the level of a hog, with the accompanying atrophy of finer elements, he is no more evil than a hog. The evil of the final degradation lies in the comparison of what is with what might have been."[14] There is no "evil in itself" in either instance. The intrusion is bad only because it arouses a furious or dogmatic rejection. But this rejection may be provisional. If the inhibition happens to be lifted, what had been greeted as an intolerable contradiction may perhaps be transformed into contrast. Such a possibility, however, is not programmed in advance, nor may it be deduced from a position of transcendence. For Whitehead as for William James, what is required is to think in its presence. What is needed is a mode of thinking that does not allow one to transcend the era or to see ahead and nevertheless refuses to ratify "by right" what at the time seems excluded or impossible.

Whitehead introduces the trick of evil precisely where we might feel tempted to adopt the venerable figure of the "truth that hurts": the truth that claims recognition through the difficulty of its acceptance and the unhappy reactions it provokes. So doing, he strips truth of all heroism, of any claim to be the bearer of a radical rupture, which would explain its rejection. Whitehead evokes the trick of evil to place this rejection under the aegis of "we do not know." We do not know whether, if it had been presented otherwise, what was proposed might have proved somewhat disturbing, but not threatening. The phrase "we do not know" is that of a mathematician, for whom a contradiction is always relative to the manner in which its terms are defined. A contradiction is liable to vanish if a definition is modified, which typically means that a part of what this definition omitted gets explained in the form of a restrictive condition of its validity. What the trick of evil represses, inhibits, and defers is the possibility of an inflection of the relation that a state of affairs maintains with what justifies it, which makes claims become stringent and uncompromising.

To place the science wars under the aegis of the trick of evil implies that there was nothing necessary about them. They did not simply translate an already existing and insurmountable conflict between critical thinkers and scientific practitioners. Critical thinkers did not in fact need to repeat the skeptical gesture of Hume and, doing so, mobilize scientists around the affirmation of a bifurcated nature. They did not need to define reality as mute, unable to provide justification for its interpretation but at best providing some constraints on interpretative freedom. In fact, many of them were already studying science "as it is done," in the field and through archives. Insightful new accounts of the sciences had already been produced that brought active, enterprising, and fully equipped scientists to the fore instead of staging a subject of knowledge interrogating itself about the legitimacy of its interpretation of facts. As for scientists themselves, they had been dealing for ages with highly constructed facts, obtained through what their instruments gave them the power to observe, instead of facts obtained from perception. To evoke Whitehead's turn of phrase, to obtain such highly constructed facts presupposes "a terrible history of complex activity." Generations of researchers and technicians developed and tested apparatuses (*dispositifs*) that were later transformed into a variety of means for making new facts possible. In short, the ancient bifurcation between primary and secondary qualities mobilized by scientists feeling under critical attack seems far off the mark.

In my opinion, the season was wrong because the theme of bifurcation had not lost any of its power over scientists and over critical thinkers. The sciences, for all their proliferating diversity, now found unity through a common claim. They reproduced the disqualification of nonscientific claims by endorsing the bifurcation of nature to relegate them to the realm of mere appearances.

Today the institution called Science systematically produces bifurcation in everything it touches. At every hand, it opposes the objectivity of facts to the subjectivity of opinions. The unity of Science is obviously a façade. A highly respected physicist will have nothing but scorn for the "facts" of a psychologist. However, once the epistemological claim to a form of knowledge that owes its authority to objective facts becomes common property, Science becomes a predatory machine. It arms the institution against whatever it calls opinion. As a corollary, scientists must make common cause because they are all contributing to the advance of objective knowledge. It is surely because their target was the authority claimed by Science that critical thinkers came to occupy the antagonistic position offered to them by the bifurcation: intransigent skepticism. Their will to demystify could not be bothered by such secondary questions as the difference between the objective measurement of radioactivity and the objective measurement of human intelligence, IQ.

The trick of evil, if such there was, lay in the critics' choice of epistemological terrain. Although these critical thinkers had learned to tell complex stories about the sciences, they gave to these stories a simple vocation, to demystify the power of facts. Their stories always returned to the same conclusion: facts are in themselves incapable of producing agreement among scientists; only after scientists reach an agreement by other means do the facts acquire objective meaning. If critical thinkers had focused on complicating the claims of Science, they might have activated critical thinking about the diversity of facts laying claim to objectivity. They might thus have called attention to the diverse stakes arising when diverse facts take on authority, and to the diverse stakeholders who stand to gain (*cui bono?*). Instead, they created a common cause. Speaking as a "we," scientists who did not have all that much in common unified in defense of science under attack.

I previously introduced the example of the measurement of radioactivity in contrast with the measurement of IQ because the latter has been a target for a number of scientists belonging to the Science for the People movement. This movement contested forms of knowledge it considered to be vectors of inequality and discrimination. Among them, the measurement of IQ was denounced as scientifically destitute and politically malignant because it was founded on school-type performances and translated sociocultural differences into intrinsic attributes of a person. Hilary Rose, a biologist involved in this movement, strove to ally herself with critical thinkers. She speaks of her uneasiness when she realized that they were not prepared to criticize IQ measurement as bad science because they did not feel there were any "good" sciences.[15] They looked on her with the ironic indulgence suited to dealing with scientists with good intentions but desperately naïve. They themselves would not be dupes. They would get to the root of the problem, which they located, once again as always, in the scientists' claim that nature had the power to produce agreement when in fact scientists relied on the power of social constructions to produce agreement.

On the whole, social studies of science and technology today have abandoned the vocation of demystification. Still, knowing how to speak about scientific facts remains an urgent question. The terrain is not the same as it was some twenty years ago at the height of the academic war around sciences. The trick of evil today could reside in the temptation to consider as a consequence of that war that those who find facts inconvenient now feel free to attack or deny them. The temptation is to reinforce the sacred union around scientific facts, to defend them whatever they may be, as innocent victims of obscurantism. To resist this temptation, we must recall that there is nothing innocent about the institution called Science. No one will get me to march in the streets in defense of Science. It is this institution that confuses facts worthy of being defended with impoverished facts deriving nourishment from the sempiternal opposition that perpetuates the bifurcation: "You believe; we know."

In contrast, anti-GMO activism is a profound source of inspiration, for activists learned how to distinguish between GMOs in the laboratory and GMOs as an agricultural innovation. They understood that a fact may be solid but only from the point of view of trials that put its solidity to the test. It remains mute about everything that these laboratory trials demand to omit. In the case of IQ, such understanding is beside the point because, in this case, the *will* to omit is the very point of the measurement. The fact harkens back to its conception, to the very project of making intelligence into a trait that is objective in the sense of being independent of life histories and the inequalities marking those histories. Is it any surprise that the result is what Stephen J. Gould called a "history of mismeasurement"?[16] In the case of the genetic modification of a living being, omission is not a project, but the price for the effective verification of the modification. This is why the obtained facts have no legitimate claim to mute the questions associated with the consequences of this innovation beyond the laboratory. We need not reduce their objectivity to an imaginary social construction. We need only emphasize that the objective definition of the modification is situated. It has been acquired or obtained in a particular territory, and is intrinsically precarious when it leaves this territory. The only objective definitions that escape this precariousness are those that have been designed to rule on any territory whatsoever, like the IQ definitions. Objective definitions then occupy territory, as it were, and like an occupying army, demand silence and submission.

We need to avoid the temptation to conflate these two kinds of objectivity, the objectivity obtained in the laboratory and the objectivity identified with the imperative to silence so-called subjective judgments. A highly significant operation is indicated in the use of the same word in both instances: the question of bifurcation is then translated from the philosophical register into a political register. It is thus not surprising that the philosophical bifurcation has survived so long. Philosophy does not bifurcate nature at all; it bifurcates the value of forms of knowledge. It is with its passage into a political register that the bifurcation begins to operate as a veritable machine of government, distributing responsibilities in a binary and asymmetrical mode. On the one hand, the ceaseless bustle of beliefs and value judgments are considered arbitrary, for ultimately, humans alone are deemed responsible for them. On the other hand, objective definitions are attributed the power to bring humans into agreement, or if not, the power to silence them. Everywhere the same imperative prevails, which puts Science in the service of public order: please provide us with the facts granting authority to your definition.

The Importance of Facts

Resisting the bifurcation of nature nonetheless demands more than the denunciation of pseudoscientific facts and the defense of true facts obtained through free and disinterested research. For Whitehead, it is a matter of resisting the idea, propagated since Galileo ("and yet it moves"), that facts obtained through science are free of values, and that it is how they can arbitrate between humans torn asunder by their conflicting values. Facts would vanquish opinion, because they are what they are regardless of what we may think about it.

For Whitehead, facts in the concrete sense of the term, in the sense of "it happened," are what we must live and think with. Yet even though such facts inexorably are what they are, their value is certainly not that of impartial judge. Facts in themselves do not have a privileged relationship to knowledge. A fact matters not only because it has happened and stubbornly leaves its mark on the future, but also because it did not have to happen, or not like that. One of Whitehead's favorite examples makes the point clearly: "Napoleon was defeated at the battle of Waterloo." Countless books explore the whys and wherefores of this fact, while others speculate about alternative histories in which Napoleon would have triumphed. The fact of Napoleon's defeat is important because it happened and might not have.

The death of a tightrope walker is a tragic fact. The fact is not that his body, like any other heavy body, fell to the ground. This fact recalls the terrible moment when he lost his balance. Such a moment can make us feel what Whitehead calls a fact "in the concrete sense." *This* fact affirms its individuality because it could possibly not have happened and yet it did, making an indelible impression on those who witnessed it. Facts that arbitrate in the world of sciences, however, are never individual. They arbitrate only because they are defined by the stakes of the situation that they arbitrate, and these stakes are general, depriving the situation of its felt individuality (which may then be called anecdotal).

As soon as we confuse the concrete individual fact and the fact that stands as proof, we fall under the sway of what Whitehead calls the myth of "finite" facts. We claim that facts in general can be isolated and defined, in manner of facts that prove. The myth of finite facts asks us to take as concrete what should demand the vigilance that we need to bring to our modes of abstraction. It thus leads to the "fallacy of misplaced concreteness."[17] But it also bears witness to the importance that the fact that proves has for us. We would wish to consider facts as divorced from values. "And yet the notion of importance is like nature itself: Expel it with a pitchfork, and it ever returns."[18] When we affirm that facts are impartial arbitrators, foreign to the values and significations that have importance for us, we are attesting to the value that renders so important this power of arbitration.[19]

Whitehead does not practice irony: he does not endeavor to make those who swear by facts confess that their facts are factious. When he highlights how facts owe their definition to the importance they take on for those who define them, he strives not to diminish facts, but to activate the vigilance needed to resist the myth of finite facts. What served as proof when dealing with the movement of celestial bodies in an ideally rarefied milieu must not be generalized as a method.

To be sure, to consider a fact to be purely and simply itself and definable as such is legitimate for certain specialized practices. It is the very stake which defines success in experimental sciences. It is the very condition for the exercise of mathematics and logic. It is what is produced by the verdict in law. But the roles facts play do not point toward a role that would reunite them, which the myth would come to generalize. The logical fact that Socrates is a man and the fact upheld by the verdict of the tribunal that will determine the guilt of Socrates have nothing in common.[20] As for the experimental fact that the neutrino has mass, which resolved an anomaly that had plagued physicists for years and was celebrated by a Nobel prize,[21] I side here with Steven Weinberg: "To tell a physicist that the laws of nature are not explanations of natural phenomena is like telling a tiger stalking prey that all flesh is grass."[22] To be able to claim as a fact that neutrinos have mass exemplifies the success that matters to experimenters.

To be sure, the neutrinos did not directly provide evidence of their mass, and the experimenters never directly experienced this mass as the tiger feels its jaws close on its prey, warm blood filling its mouth. But if we may speak in this instance of "obtained" objectivity, it is because for experimenters to obtain objectivity is not the same as what critical thinkers call social construction. Such an obtaining evidently supposes a number of such constructions, and critics may track, for instance, how it was first necessary to persuade fund-granting institutions and colleagues in competition for the same funds. Obtaining funds for the extraordinarily costly and sophisticated apparatus needed to prove that the neutrino had mass indeed required a social history of negotiations, alliances, and argumentation that contributed to making the presupposed mass of the neutrino something of importance to "anyone." Ascertaining the possibility that the neutrino has a mass became so desirable and so eagerly awaited by everyone involved that it was in the interest of each of them to find proof of its hypothetical existence. But this is also where experimenters may take mortal offense and feel ready to go to war. Interest does not explain that it indeed has a mass.

What mattered to experimenters is that the exceedingly costly and sophisticated apparatus for which they made arrangements allowed them to create an unprecedented relationship with something they call neutrinos. That this bringing into relationship (*mise en rapport*) may permit them, depending on the outcome, to attribute mass to neutrinos, has for the experimenters a very particular meaning: in case of success, no one should be able to assert that the attribution of mass is merely the product of human interpretation or of agreement among humans. Obtaining such a success signifies that, for experimenters, any social construction that brought the sophisticated apparatus into being is only a means. It has value only through the outcome, the conclusions, the answer to the question obtained.

This is why saying to experimenters, "this neutrino henceforth endowed with mass is still and always nothing but an abstraction for which humans are responsible," is like saying "all flesh is grass." The passion of the tiger for flesh is correlated with the possibility of capture and the risk of failure. Grass does not flee before grazers. The mode of abstraction invented in the laboratory has the singular value of supposing that the attempt at capture may fail. Neutrinos must be rendered capable of ruining human expectations. The specialists' concern over the difference between a neutrino with mass and without mass, then, is a case suited neither to logicomathematical abstractions nor to juridical abstractions (if accusations are judged insufficient, a different fact will be established: Socrates is not guilty). In this respect, the experimental fact is not *exemplary*, but *exceptional*. This is why experimenters are known to dance in their laboratories, which is not the case with logicians and jurors. Still, they remain aware that their success risks failing to interest anyone but their concerned colleagues.

What scientists call disinterested research thus does not at all mean research that ignores the set of interests that it may arouse. Yet it demands that these interests remain suspended until success, of which they alone may be judge. They alone are called on, and they alone are able to put to the test what is presented as an experimental fact, to determine whether success has really occurred. Is it possible in such a case to attribute responsibility to the interrogated phenomenon for the answer obtained? Is it possible to assert that the phenomenon authorizes the scientist to speak in its name?[23] Such questions imply that the experimental mode of abstraction, as carried out through the apparatus, needs to confront the possibility that it might have omitted something of importance, something whose omission may imply that the fact obtained does not authorize anything at all.

The criterion of experimental success explodes the myth of the isolated fact, making the possibility of isolation a rare success, potentially costly and, more than anything, situated. The fact, which may claim to be purely and simply itself, and capable as such of imposing itself against all subjective interpretation, is a fact obtained in the laboratory. As such, its capacity to impose itself is limited to the collective of concerned experimenters. It does not impose itself against subjective interpretations, but against competent objections on the part of those whose responsibility it is to voice them: "Take care, your apparatus has overlooked this possibility, and if you do not succeed in ruling it out, it may all be wrong!" The experimental sciences have been able to subscribe to bifurcation, but also, thanks to their successes, to expand the manner in which allegedly objective nature is populated—with the caveat that speaking here of nature implies a propaganda operation that often seeks justification in the thesis that people are incapable of understanding the requirements and obligations associated with experimental success.

The objectivity thus obtained does not tolerate irony (this is but a construction), and yet, if those who obtain it had confidence in the capacity of people to get interested in what matters for them, it might open itself both to humor (there exist worlds beyond the laboratory!) and to the vigilance it requires. As we have seen with GMOs, what the protected and well-equipped environment of the laboratory permits experimenters to omit may take on great importance upon a single step beyond its wall. Put another way, experimenters might behave in a "civilized' manner." They might contribute to a thoughtful culture of facts, respecting their intrinsic diversity. They might not define themselves against opinion.

In contrast, what we call Science today has claimed for itself a definition of objectivity that amalgamates the idea of our forms of knowledge irresistibly advancing, like an oversized yet blandly monotonous wave, with the necessity of arbitration to ensure public order. Science insists on the bifurcation, even if it strips away, to the point of absurdity, everything that makes a situation matter to us. For Science, the only question of any value is: "How can we redefine a situation in a manner that will bend it to an objective definition?" Worse, Science today demands that experimenters accept that their successes count for very little with their granting agencies, who set a premium on innovation as the bearer of growth.

Even in desperation we must resist the apocalyptic idea declaring that we now live in a post-fact era, which would also be post–common-sense because our compasses are all lost or broken. Whitehead, too, lived through an era that spelled the defeat of common sense yet proposed to weld common sense and the imagination. Today, he would surely say that it is now all the more important to learn to speak well of facts. Surely the manner in which facts have been enlisted to silence and disqualify doubts and protests does not suffice to explain what is happening today when the right "not to believe" in facts is claimed. It reminds us, however, that coming up with a "good" answer to this question may not be the point. The point may rather be to let ourselves be situated by a task. If we have to learn today how to speak well of facts, it is because we have tolerated too many myths and abuses. This is what situates us. This is what incurs our responsibility.

There is so much to do because there are so many kinds of facts! There is an entire culture of facts for us to activate, to vivify, to make matter, but also to prevent from doing harm. The fallacy of misplaced concreteness can have devastating consequences. From Alice Rivières, I learned how a fact well-established through experimental means could shatter a life. The fact concerned the genetic mutation responsible for Huntington's disease, a neuroevolutionary disease against which contemporary medicine is powerless. It presents a classic case of an effectively finite fact: the existence of a well-identified genetic mutation makes possible a perfectly reliable presymptomatic diagnosis. But the doctor who announced to her the results of her test acted as the spokesperson of the laboratory, thinking it his duty not to leave her with any illusions about what awaited her. His was an error of exceedingly misplaced concreteness. Not only is the course of the disease very unpredictable, but his mode of address also introduced a bifurcation between objective prediction and the "subjective" response to the ordeal. The doctor acted as if Alice's capacity to anticipate, prepare for, and live with what awaited her did not matter, or was none of his business. The abstract truth-telling of the test was transformed into a staggering verdict, transforming the landscape of possibilities into a wasteland.[24]

It is an entirely other matter with the facts obtained through experimental science but produced by the functioning of detector apparatuses such as Geiger counters. The detected fact is really and truly finite, but in this case, it does not arbitrate on anything at all. It signals something of importance but does not define the situation. For better or worse, in the zones affected by the catastrophes of Chernobyl and Fukushima, Geiger counters are now part of the everyday life of inhabitants, like a sensory organ, rendering perceptible and localizable what would otherwise be a diffuse and omnipresent threat. The finitude of the fact of detection does not separate Science and opinion. It adds, for better *and* worse, to the entanglement of ingredients that enter into a concrete situation and demand attention.

The same is not true for the fact called "global warming." This fact cannot be separated from a political initiative, the creation of the International Panel on Climate Change (IPCC), whose mission was to evaluate information for governments and populations to better understand the risks connected to climate change of human origin. It is thus a deliberate attempt by scientists to get out of laboratories and convince people that the threat really exists, with consequences demanding action. Those who deny climate change insist the IPCC does politics, not science. This implies that scientists would create arguments out of nothing, calculated simply to promote a political position. The deniers are right in one and only one respect. The climatologists involved in the IPCC know their climatological models are designed first and foremost for the public and politicians, not to generate disinterested knowledge. Their observations and models have allowed them to speak effectively about global warming as a fact, showing how the ongoing increase in gas emissions is responsible for the greenhouse effect. But the success of this fact has not inspired researchers to dance in their laboratories. In addition, unlike those scientists who put on display the grand opportunities for innovation made possible through their research, the IPCC scientists know they must convince skeptical, reluctant interlocutors. Stephen Schneider, who was on the front line until his death in 2010, often found himself consciously caught between his scientific ethic (to include doubts, precautions, and conditionals with respect to facts) and his loyalty to the earth and its inhabitants.[25] The earth itself certainly had *in fine* the power to convince even the most recalcitrant, but could do so only through the sort of concrete, dreadful facts that had to be avoided. Here we feel the cruel lack of a culture of facts, the understanding of what one may and may not ask of climate models and the acceptance that they will not tell us how to respond to the questions they impose on us.

Finally come all those facts referenced in our regulations that allow or forbid, that arbitrate between divergent interests, and that intervene in concrete domains of practice that should cause our modes of abstraction to stammer. Contrary to experimental facts, these facts do not herald the realization of a possibility, the happy ending of a suspense. They belong instead to the regime of governmental necessity. To make arbitration possible, facts are essential— facts on which stakeholders come to agree and not facts that allow them to reach an agreement. Such facts may be called "conventional" facts. This appellation is not intended to denigrate conventions. Neither does it oppose conventions to a truth superior to them. A convention derives its value from the agreement it institutes and from the care devoted to maintaining the quality of this agreement. This is why, when it comes to conventions, the bifurcation is truly poisonous. The bifurcation compounds conventional agreements with the theme of the objective fact. It thus confers on the conventional agreement the authority to rule out what is deemed merely subjective. The care required to maintain conventions is forgotten. Forgotten too is the rare and highly selective character of the facts defined as objective through experimental sciences. The possibility of an objective definition is no longer an event. It is required in order to mute opinion. Compounded with objectivity, conventional agreement is now used against an opinion that is defined as dangerously suggestible, always ready to succumb to panic, or to side with charlatans.

The Art of Conventions

Whitehead might be faulted for considering modernity in terms of its symptoms. His work helps us draw up a clinical picture without identifying the disease. This may be for the best. After all, analyses of the disease, commonly called capitalism, have relentlessly gained in complexity, and even the most famous run the risk of being outdated.[26] Karl Marx understood capitalism in terms of class struggle. And so he also imagined the possibility of capitalism producing within itself the force capable of overcoming it. Today such a perspective is liable to be accused of validating implacable modernization. Only reactionary nostalgia for the past would cause lingering over what has been destroyed, the interwoven ways of living and dwelling, human and nonhuman.

Whitehead's attention to the bifurcation of nature brings something else into play: the defeat of commonsense, the misuse of facts, and also, I would now add, the transformation of convention into an instrument for the maintenance of public order. It may be why his thought is contemporary to us. When activists today declare, "we do not defend nature, we are nature defending itself," they may not be criticizing the bifurcation of nature as such, but neither are they activating the trick of evil. Their stance has got the power to make hesitate those who suddenly realize how much the question of nature has changed.

The activists call on a nature to which humans belong. They thus abandon what might have been expected to unite all of us, be it the cult of an objective knowledge of nature or the call for everyone to respect its rights and accept our duty to protect them. At the same time, they uphold a sense of proximity to peoples for whom what we call nature has never been considered in terms of objective knowledge nor of moral righteousness but in terms of attentiveness, care, and prudence, as well as fear and gratitude. The activists' sense of proximity with Indigenous peoples is not limited to alliance with their struggles. It extends to learning from Indigenous peoples who are struggling today what has sustained their ongoing resistance to modernization. This sense of proximity thus situates those who experience it in a mode that resonates with the question of this book. Are they contributing to a "becoming civilized" of modernity, or to a becoming civilized marking the end of modernity?

Needless to say, activists do not have the answer, and whatever their answer might be, it would have nothing to do with the pet themes of academics who expound on our so-called era, be it postmodern or posthuman. The answer belongs, in fact, to those who in the future, if future there be, will recount to their children what made this future possible.

While we do not know how these stories will be told or whether they will narrate the becoming civilized of our civilization or its end, they matter to us because they provide imaginative nourishment for our inquiry without answer. Some of these stories might well look back on how we gave disproportionate importance to facts that could lay claim to a strange power to bring all parties concerned into agreement. They might dramatize the corresponding absence of a culture of facts, of those conventional facts to which we must give authority in order to stop disputes. These stories would be as strange for them as are to us the memories of a time when confessions extracted through torture were held to be truthful.

Asking what memories we will leave is not an epistemological question. The bifurcation of nature explains neither the role of modern conventional definitions to ensure public order nor the privilege of successfully obtained objective definitions to establish an agreement. Dividing the territory between them, the two kinds of definitions work in a mode that renders the art of reaching an agreement null and void at the outset. But an art of agreement has been cultivated throughout the world, notably among the Indigenous peoples of the American continent and in Africa, where Portuguese colonizers christened it *palavra*. The term *palavra*, or "words" in Portuguese, speaks directly to what surprised the Portuguese colonizers: the interminable exchanges of words, which often felt idle, that the African peoples imposed on them in order to reach the least agreement. If there is proximity today between activist movements and Indigenous peoples, it arises especially through the necessity that is felt to work for the resurgence of such arts.

We do not know what happened in those meetings in which African peoples encountered those who were to "civilize" them. I am among those who have had the opportunity today to participate in this kind of exchange in the African milieu, and I have experienced exchanges that have nothing idle about them. But they oblige acceptance of constraints unusual for lettered Europeans. The exchange is not a matter of democratic debate subordinated to an abstract principle of equality, because those who participate in it are enrolled as elders. Nonetheless, this assumed role does not put them in a position to make decisions. An elder cannot be contradicted, but she herself is obliged by her role to draw from her experience syntaxes, rhythms, and manners of speaking that do not address the possibility of a contradiction. She will not put forth an "I" with intention to defend her reasons. Her speech brings into existence the impersonal experience that makes an elder of her. Each word she speaks must be produced so as to express a given dimension of the question that brings everyone together, even though the question brings them together because there are hesitations, points of divergence, and risk of conflict.

We can speak, with regard to this apparatus, of an art of convention. Certain peoples summon beings in whose presence they are to speak (and others not), but the principal trait in common, it seems to me, is the efficacy of that which, by reference to what we have inherited from the Greeks as a natural form of discussion, should be called "artifice." In effect, this apparatus has as its effect to impede what is required for us as the very principle of free deliberation: each has the right to express herself, reasons come into conflict, and victory must be given to one of the rivals.[27]

To speak of artifice is to avoid giving priority to the question of beliefs that we would no longer share. Thus, I speak of an art, of practices whose existence depends above all on their efficacy. The role of elder belongs to this art, because it contributes to a mode of meeting in which the manner of listening to others will not involve preparations for a counterargument or an act of interpretation seeking intentions behind what is said. Each speech participates in a long and often apparently repetitive process, and if the process generates decision, no one will be able to appropriate the decision, and nothing will be able to guarantee that it is the best possible decision. The decision has received *its* reasons.

We may expect the modern proponent to object: "All these histories of efficacy and generativity are first and foremost a manifestation of your credulity: nothing guarantees that the agreement will amount to much more than a clandestine play of influences, rhetorical effects, and group conformity. Besides, those who meet all know they are supposed to come to some kind of agreement. Even if they are not conscious of it, each is already prepared to let themselves be influenced by the others." Such an objection gives expression to another facet of the bifurcation of nature. Apparently, for the modern proponent, the fundamental danger is being a dupe, of not seeing that influences are at play in the decision. Or, in other terms, the danger of being duped is promoted to eliminate possibilities opened by the efficacy of *palavra*.

Those who practice nonviolent direct action today, however, have need of the art of consensus decision making, because they know such a practice may help them to resist the test that awaits them. They know that they will confront provocations and violence that will make it difficult for them to stick to their commitment of not responding in kind. They are keenly aware that, if influence games, conformism, and submission to a dominant rhetorical stance have determined their choice of action, it may not hold up during the action, and during the aftermath when everything will be done to divide them. This is why they have invented roles and constraints whose efficacy is formulated explicitly. Decisions are made with respect to a situation that must not be reduced to isolated facts that would provide the argument for a thesis or objections to the thesis. The situation must be unfolded out as concretely as possible, and to do so demands that many divergent voices be heard, not as contradicting each other, but rather as expressing its many different dimensions.[28] Arriving at a decision implies the efficacy of an artifice. Refraining from bending the situation to an argument goes against our ingrained habits. It constrains one not to hear the others in the usual manner: not to object to them and defend one's position. The efficacy of this apparatus consists in giving the situation the power to make everyone hesitate, to attune diverse reasons, and to generate statements that lose the power of contradicting others. If a decision is obtained, it will express the way in which the situation has gained the power of making sense in common. The decision is something obtained, not the result of artifice. Artifice is necessary in its way to ensure vigilance toward our modes of abstraction. It counteracts the tendency to define a situation in terms that insist on the importance of knowing who is right. Artifice allows a decision to be obtained without anyone being the victor.

The culture of what I will call "generative apparatuses" is central to practices of direct democracy. As activists acquire skills in these practices, they are discovering that many of the Indigenous peoples knew how to preserve them despite the devastating effects of colonization. These ways of making sense in common are today making possible actions in common against the devastation pursued in the name of economic growth. If future there be, these apparatuses may then appear as obvious and multifarious as the situations a collective may confront. Will they have contributed to a process in which modernity civilized itself? Or will modernity be relegated to the past because it deemed such apparatuses impossible or deceptive?

In any event, we now have a better understanding of the modern bifurcation between facts (neutral and impartial) and values (synonymous with bias). When the moderns turned to free discussion to determine what actions to take, they transformed the difficult art of reaching an agreement into a two-step process. They needed the authority of facts to define what is, and then would freely argue about what should be. They thus took for granted what, in the case of experimental facts, is artfully obtained.

In a way, experimental facts indeed depend on a rather particular art of agreement. As we have seen, they exist only because they have obtained the agreement of those who are united by the possibility of such an agreement. Competent colleagues do not wish to come to an agreement, but to be forced to agree by the experimental situation. Competent objections are thus welcome, as they are needed to unfold the requirements of a successful experimental outcome. Such an agreement is generated by the power given to the experimental situation to make experimenters think and imagine together, but its scope is intrinsically precarious, because it depends on the exacting demands defining such a situation and the correlatively involved competency.

In contrast, the activist's art of consensus or the *palavra* do not hinge on the power of an obtained well-defined fact to impose agreement. They rather aim for the creation of what Whitehead calls the experience of an individual concrete fact, a fact that does not attest to reasons more general than it is, that thus cannot be abstracted from the process from which it issues, reducing it to a mere means. The agreement thus created attains an individuality that is inseparable from a process of composition allowing new possibilities for speaking and feeling to emerge. It makes possible the transformation of antagonistic reasons into contrasts that matter. It separates words from the purpose of conveying intentions and significations that should be accepted by everyone. It allows for as many expressions of a situation as there are aspects that come to matter. Yet it does not claim any ability to define the situation. A common sense may come into existence, a consensus we might even say, but not in the sense of a unanimous agreement or a unique and totalizing voice. The situation makes sense in common. It generates a feeling together, each one in their manner, but with others and thanks to others.

The success associated with the experimental fact and the success sought through generative apparatuses are of two kinds, to be contrasted but not opposed to one another. I propose, in order to contrast them, returning to the divergence between two adventures whose common trait is that their successes refuse to be dismembered in terms of mutually exclusive responsibilities. In fact, when the experimenter dares to say nature has spoken, responsibility is attributable neither to nature nor to the human mind. What is celebrated is a rare event—that which the human interrogated has agreed to assume the role proposed for it by the experimental apparatus conceived by the human. But the second kind of success is an event too. An agreement has been created that is capable of making sense in common with respect to a problematic situation, and the reason for the agreement can be attributed neither to the situation nor to the manner in which humans have interpreted it.

There is nothing exceptional about such events. Take the apparatus put in place by Citizens' Conventions, which gather around questions bearing on technoindustrial innovations.[29] Such an apparatus gives some sense of what a culture of conventional facts might be. Citizens' Conventions take on conventional facts that are today omnipresent yet poorly nurtured, left to obscure meetings among experts who represent divergent interests. They make possible for "ordinary" people, previously "ignorant," to think and learn together how to listen to debates among experts without desperately seeking the right answer or the legitimate position. They acquire the skills and imagination to formulate questions, propositions, and objections whose pertinence and clarity are able to make the experts stammer.

This apparatus calls on an ancient artifice, the drawing of lots. Its particular efficacy is to select participants independently of their particular merits in regard to the problematic situation. In this way, it introduces into the situation the possibility of a relationship that we may call disinterested in the sense that it does not privilege a specific concern that would in turn correspond to a particular competency on the part of the person selected, justifying her selection. The participant knows she is elected only in her capacity as anyone whosoever. She knows that her identity and personal merits are not in play. Her legitimacy is not the point of reference and will not have to be demonstrated. What is effectively at stake is the quality of her participation in the creation of the collective space of thought demanded by the situation to be explored with others.

Participating in the drawing of lots implies a prior acceptance of a role that makes demands: for several weekends, the elected familiarize themselves with the dossier, posing questions of experts, which may lead to calling in additional experts. The crucial point is that they create a relationship with expert forms of knowledge in a manner that breaks with "natural" pedagogical procedures which advance from ignorance to knowledge already mapped in advance. What makes them collectively intelligent is the unmapped question bringing them together, which no single form of expertise would suffice to define. While experts approach the situation in terms that validate their respective forms of knowledge, the group operates by problematizing those forms of knowledge. The group plunges the known into the possible, activating unknowns that situate the known in the mode of "yes but." It densifies and intensifies the question. It does not dismiss anything, yet troubles the power of expert modes of abstraction to define the situation. The group spurs the emergence of values that till then have been only a stammering within speech. To evoke Donna Haraway, the capacity to accept "staying with the trouble" becomes the object of an affirmative realization—*sheer disclosure*.[30]

The creation of a group in which common sense and imagination may become welded together is remarkable in itself. But equally remarkable is the intensity experienced by those who participate in it. They find themselves capable of actively taking part in the collective exploration of a problem that was supposedly beyond them. The feelings of joy and new esteem for self and others, as well as the desire to revive the experience, are diametrically opposed to the heinous feeling of humiliation that I have come to associate with the unflagging support Donald Trump elicits from his supporters. But this joy is fragile: it may give way to disenchanted cynicism if the group feels betrayed. A sense of betrayal could set in if the advice offered remains without consequence, as is often the case today, or if it turns out that the terms for the problem posed to citizens had in effect been decided in high places already.

While it is the efficacy of such generative apparatuses to produce intelligence in the sense of a capacity to think, imagine, and explore what matters, intelligence is not an attribute of the apparatus. It is obtained. Just as it is routinely destroyed by other kinds of apparatuses that, in one way or another, presume ignorance, manufacture fears of being mistaken and a lack of confidence, or create a situation in which each person wishes her interest to prevail and remains suspicious of others' interests. Generative apparatuses testify that intelligence can be regenerated where brooding has not been eradicated, where the refusal to think and to imagine has not prevailed, and where joy has not given way to hatred of the world.

Whitehead's remarks on the solemnity of the world help to situate the success of generative apparatuses. "All forms of realization express some aspect of finitude. Such a form expresses its nature as being this and not that. In other words, it expresses exclusion; and exclusion means finitude. The full solemnity of the world arises from the sense of positive achievement within the finite, combined with the sense of modes of infinitude stretching beyond each finite fact."³¹ What a generative apparatus obtains comes without guarantee. Because it entails the realization of an individual concrete fact that determines itself in this way and not any other, its reason cannot be grounded on something more general. The importance of the generative apparatus lies not only in the quality of what they allow us to obtain. It also comes of the keen sense of positive achievement arising from the obtaining. This manner of obtaining generative togetherness enables the experience of how not to "deface the value experience which is the very essence of the universe."³²

{ 3 }

A Coherence to Be Created

Common sense, in Whitehead's account, entails brooding over aspects of existence. Common sense does not wait for someone to tell it how to think when it takes to brooding, still less does it wait for someone to choose which questions it has the right to pose and which ones ought to be renounced because the answers would outstrip what can be known. Brooding is characterized instead by a silent insistence, a "yes, but." Yes, we know very well that our modes of understanding are finite in character. But we refuse to accept that what we nevertheless do know is empty of meaning, even if it is outside the boundary defining what we may legitimately claim to know.

We Require to Understand

This form of knowledge, in its silent insistence, concerns "aspects of existence." Existence here is not an object of knowledge. Existence is something we enjoy, something we apprehend concretely as enjoyment. Aspects of existence are thus not readily defined by reflexive or discursive consciousness. On the contrary, they often become conscious only with effort and hesitation, when we wish to push them forward and put them into words. When the words they require are assembled, they often seem to be mutually exclusive. This difficulty characterizes the series of "we require to understand" that appears in *Modes of Thought*.

We require to understand how the unity of the universe requires its multiplicity.

We require to understand how infinitude requires the finite.

We require to understand how each immediately present existence requires its past, antecedent to itself, and requires its future, an essential factor in its own existence.

We require to understand how mere matter-of-fact refuses to be deprived of its relevance to potentialities beyond its own actuality of realization.[1]

In the next section, I will return to each term in this series, not separately, but as so many aspects of the metaphysics of Whitehead. Understanding for Whitehead does mean becoming capable of explaining something to someone, or arguing and proving something. Understanding is close to aesthetic enjoyment, wherein the effect produced by the totality precedes the discrimination of details. Details assert themselves subsequently as reasons for the effect. Whitehead writes, "in aesthetics, there is a totality disclosing its component parts."[2] In contrast, a metaphysical system is not an aesthetic work because an aesthetic work is, as Whitehead underscores, a closed fact that "produces totality." To awaken the sense of "the solemnity of the world [that] arises from the sense of positive achievement within the finite,"[3] Whitehead's system deploys aspects of existence, but not as parts of a totality that would give them their meaning and necessity. Aspects of existence as such participate in the solemnity of the world. Whitehead's system aims to give to the individual concrete fact the power of making itself felt as an achievement at once for itself and for the world.

To speak of achievement is to speak of value. Recall that the perfect knowledge offered by Laplace's demon corresponds to a reduction of the world into a succession of states defined as equivalent. If everything has the same value, nothing has value. It is only humans, then, who confer values on an indifferent nature. I begin with the question of value because everything began there for Whitehead, with the claim that values are intrinsic to what scientists study, the order of nature. This claim mobilizes the order of nature against the bifurcated nature he refuses.

As early as *Science and the Modern World*, Whitehead came to the conclusion that the first victim of the bifurcation of nature is biology. Because the bifurcation separates the physical world, with its universal laws, from the appearances presupposed by human life, it summons biology to explain how living beings can emerge from an indifferent law-abiding reality. Living beings are not "given" as electrons, protons, and other physical entities are supposed to be. They are "enduring" in the sense of a positive achievement. The biologist does not call this reality of living beings into question, because this achievement situates her as well. She seeks to understand how it holds together. Additional analyses produce additional surprises. She learns to pose questions that do not verify a hypothesis derived from theory. Her question comes from what she observes and what intrigues her—how does it achieve that? Such a question is always addressed at *that* particular being, which, affecting its milieu and affected by it, succeeds in maintaining its existence in a mode that is never general.

The biologist must learn from each particular being, because the pathways of deduction are impracticable here. Every relationship between what is achieved and how it holds together must be diagnosed on its own terms, for it might have been otherwise. For biologists, the aura of "what might have been" always hovers over "what is." This is how many of them come to experience of the solemnity of the world.

Whitehead christened "organism" any entity whose existence required maintenance to exist. The reality of an organism is an ongoing, enduring realization that depends on the patience of its environment with respect to what this realization demands of it. Because this realization always has a determined form, it is in itself the achievement of a value. Its value comes of its success in obstinately holding together and retaining itself in *this* way, its own way.

> The salvation of reality is its obstinate, irreducible, matter-of-fact entities, which are limited to be no other than themselves. Neither science, nor art, nor creative action can tear itself away from obstinate, irreducible, limited facts. The endurance of things has its significance in the self-retention of that which imposes itself as a definite attainment for its own sake. That

which endures is limited, obstructive, intolerant, infecting its environment with its own aspects. But it is not self-sufficient. The aspects of all things enter into its very nature. It is only itself as drawing together into its own limitation the larger whole in which it finds itself.[4]

What saves reality from becoming illusory are the obstinate facts capable of obstructing, of ruthlessly interrupting the flight of theoretical reasoning. Reality is thus saved from its reduction to a theoretical question: are there primary qualities, independent of the human mind? The Thracian girl laughs at Thales, who, head in the stars, has forgotten that his heavy body requires the earth to support it. This is the reality invoked in the last instance by common sense, when it is confronted by a theoretician who aims to prove that its reasons for believing in the world do not hold. Whitehead takes the side of this reality: reality is what we collide into, what collides into us, whether we want it to or not.

To think this reality defined by and for itself, however, Whitehead eschews what surely is the first thing to come to the physicist's mind: the weighty and inert stone whose fall might crush Thales's brain or mute the Thracian girl's laugh. The stone inspires visions of self-sufficiency, of something that holds together by itself with no special need of an environment. As for its fall, it inspires visions of inexorable submission to an impartial, indifferent law. The stone is directly connected to the reality of physicists, a world stripped of meaning and value. Its neutral and impassive obedience encourages an image of common sense in revolt: when common sense finds all its words cut short by a pedagogic, inexorable dismissal, it may brusquely interrupt the exchange, become stone deaf, thus attesting to its irrationality.

Whitehead turns instead to the organism of biologists. Biologists, who deal with the organism, must ponder and use their imagination to figure out how the organism succeeds in holding together. Also, the organism does not interrupt indifferently. It affects or infects its environment and is affected or infected in return, each in their own way. In particular, an organism does not enter indifferently into biologists' experiments; it responds in its manner, according to its own criteria of valuation.

In *Science and the Modern World,* Whitehead proposed to make the concept of organism central to ontological questions about entities that populate this world. His proposition changes a great number of things. Physicalist ontology is connected to the idea of a world knowable, as such, independently of us. It subordinates the plurality of sciences to the imperative of recognizing that they all deal with the true texture of the world as physics defines it. In contrast, what Whitehead calls the philosophy of organism implies a positive plurality of sciences, studying the entangled texture characteristic of the different types of organisms that make up this world. To know is always to know something that endures and to learn from its capacity and manner of enduring. But to know an enduring being is also to be situated by it. A being has its way of valuing its environment, and the one who attempts to gain knowledge belongs to this environment. To know a being is thus to learn which questions are made pertinent by its way of valuing the questioning situation.

Here again, Whitehead reconnects with common sense. Just consider the success of animal stories, the interest aroused by the question of knowing what an animal, or more precisely *this* animal, is sensitive to, how it values its environment, what it makes matter. Common sense is realist in that it knows there are ways to *learn from* a fundamentally plural and entangled world. To learn: to acquaint oneself with, to track, to investigate, to taste, to test, to induce by analogy, to discover, and maybe to understand—but never in general, always relative to a question, a concern, or a situation.

For many sciences, in contrast, such a pluralism has somewhat disturbing consequences. These sciences are used to actively and unilaterally shape and value what they deal with in terms of their own requirements for objectivity. Thus, for these sciences, to learn would mean to accept the necessity of preventing their terms from unilaterally dominating the situation. This is precisely what Vinciane Despret asks of animal ethology. She stresses how researchers should learn, and are indeed in the process of learning, to pose good questions of animals, sharing situations with them in which the animals are "going about their business."[5] It is animal business, the way animals make their environment matter, that ought to situate the researcher and bring her into relation with the animal, for it allows for potential congruence between their divergent interests. By the same token, in sociology, if investigators wish to obtain genuine answers, they need to pose only those questions that actively engage the concerns of those whom they question.[6]

Whitehead often speaks of his philosophy as a philosophy of the organism, which implies that the notion of organism is not limited in scope to what he calls the order of nature. The notion of "organism" goes beyond questions about how the multiplicity of ways of enduring composes the world. It is connected with questions about what matters *for* what we address, and how. It associates the sense of worth with "the sense of existence for its own sake, of existence which is its own justification, of existence with its own character."[7] In *Modes of Thought*, Whitehead no longer cries out against the absurdity of ideologies of the bifurcation of nature, as he did in *The Concept of Nature*. The cry now obtains the full force of an affirmation that I evoked at the end of the previous chapter: "We have no right to deface the value experience which is the very essence of the universe."[8]

Dare to Speculate

For Whitehead, humans are characterized by the overweening importance and worth they confer on what is possible. Irony is wasted on them, as are attempts to bring them back to the solid ground of factuality. Even experimenters do not stick to the ground of factuality. Facts are worthy of their attention only if they attest to what matters to them: the possibility that a fact may have the power to confirm certain interpretations and to disqualify others. Nor is it any use to elevate certain possibilities to the status of ultimate horizon, which would lend its virtue to the quest for Man. Even if studying the order of nature presupposes and intensifies the sense of what is possible, it is far from exhausting what we know or what we require to understand. On the contrary, studying the order of nature entails specialized research, and the modes of abstraction needed for such research bring into play a pronounced rift. On the one hand, the researcher strives passionately to formulate pertinent questions. On the other is that which is called to become a witness about how it should be characterized, how the way it values its environment explains its behavior. At the end of *Science in the Modern World*,

Whitehead appeals to this passionate kind of special investigation in a way that initiates the shift in viewpoint that would lead to writing *Process and Reality*. "In the present chapter, and in the immediately succeeding chapter, we will forget the peculiar problems of modern science, and will put ourselves at the standpoint of a dispassionate consideration of the nature of things, antecedently to any special investigation into their details. Such a standpoint is termed 'metaphysical.'"[9]

Dispassionate is entirely at odds with indifferent, neutral, or disengaged. It corresponds rather to the engagement of the philosopher with aspects of existence that make common sense brood, aspects that modern thought has passionately dismembered into openly irreconcilable characterizations. How to reconcile a Newton, whose work demands ardent confidence in the possibility of deciphering the order of nature, with the order of nature as he has deciphered it, dominated by a mode of universal and indifferent causality and imparting no meaning to what is possible? Is it necessary then to listen to Hume, for whom this order is only a construction, projected onto a mute reality, but who must nonetheless silence the testimony of his own body about how this reality affects it? Or Kant, who relegated our joys, pains, regrets, and terrors to the web of causes, and who admitted as true value only the respect of the universal moral imperative addressed to a subject whose liberty is defined against causes?

To counter the predatory power modernity has lent to certain modes of abstraction, Whitehead engages metaphysics. His aim is to bring coherence, to make coherent what specialized, passionate investigation takes apart. Critical thought or reflexive thinking is not suited to this task. In its bid to denounce improper generalizations, critical thinking insists on finitude, but such finitude is not associated with a sense of positive achievement. Doubling it like a shadow becomes a reference to an unattainable truth, over and above any ground, over and above any milieu, a truth we have to renounce, remembering time and again that we have no access to it. The duty of interminable renouncing may act to block any call to adventure. But it also affirms, in an underhanded way, Man as exception—he who has tasted the forbidden fruit will never regain animal innocence. William James's pragmatism, which ignored this duty, was reduced to the impoverished thinking of the businessman who holds that what matters is what pays off. As for Whitehead, the mathematician, he summons us to adventure with the doctrine of evolution in its radical sense, without duty or principles. A mathematician has no other duty than the one associated with the problem she addresses. Whitehead will turn the dismemberment of our sense of existence into a call for coherence as a problem that calls for a solution. Coherence is not to be discovered beyond finitude. It will not require muzzling, limiting, or domesticating the power of our modes of abstraction. Coherence is to be created through the positive "problematization" of the predatory character of this power.

What Leibniz called his "grand moral advice" may prove pertinent here in connection to the question of the "civilization" of our modes of abstraction. Also a mathematician, Leibniz advised the one who is about to act to ask the question *dic cur hic? respice finem?*—What am I doing? What is the purpose of this?[10] Leibniz elaborates on the artifice of this question. At stake in the question is not the answer to be given. At stake is an affective or existential transformation, an enlargement of the imagination. By the terms of the Leibnizian system, no one can provide a truthful answer to such questions anyway, except God. But God, of course, does not answer any questions. No one has access to the truth of His reasons. No one has access to the why of a particular choice, nor can anyone define the purpose it serves. Adam, at the moment he took a bite of the apple, could not have said why, or to what end, because, according to the Leibnizian system, conspiring with Adam's act is the world chosen by God. The purpose of his act thus belongs only to God. The divine calculation of the best has determined the choice of the world in which Adam would eat the apple. As such, the efficacy of the question *dic cur hic?* comes from how it problematizes the general reasons we are prone to give by making the "here" matter: "Slow down, let yourself be affected by the *hic, this* situation to which you are about to respond or react in *this* way." We might restate his advice in evolutionary terms: "You may of course distinguish yourself from other terrestrial animals, such as we understand them; you may claim you have reasons to choose or act. But it is the importance you accord to your reasons that distinguishes you, not actual access to a general, non-situated reason with the power to justify yours." Problematization does not go back to something more general. It confers on the situation, always such and such a situation, the power of posing a challenge to what might seem self-evident. Problematization thinks through the milieu, so to speak, instead of asking a highly rarefied milieu to ratify a particular mode of abstraction.

It probably took mathematicians like Leibniz or Whitehead to situate problematization at an affective and existential level where it affects our relationship with our own reasons, instead of making problematization into a reflexive or critical practice. In the wake of Kant, Leibniz became the very example of speculative thinking whose temptation philosophy must henceforth resist. Whitehead, however, presents his magnum opus, *Process and Reality*, as an "essay of speculative philosophy," as if the reasons for Kant's ban were of no concern to him. Their common trait may well be that both Leibniz and Whitehead propose a system, as if they had access to the very truth of the world, to what creates its coherence beyond appearances. Yet, instead of a triumph of deductive thinking through obedience to true principles, their system is a machine to "make their readers and themselves think." The only truth for Whitehead is the solemnity of the world, but this solemnity does not traffic in principles.

Whoever tries to read *Process and Reality* either closes the book or realizes that the truth of the Whiteheadian system resides entirely in its effects, in the enlargement of the imagination it inspires. Above all, its truth has nothing to do with truth based on premises: "It must be clearly understood... that we are not arguing from well-defined premises. Philosophy is the search for premises. It is not deduction."[11] In the introduction to *Process and Reality*, Whitehead presents his conceptual schema in terms of premises for a system of which all experience would be application. Instead of something that follows from premises, however, application here must be understood as something that has the power to verify them. The system entails verification in James's sense: applications put premises to the test. Premises should never justify subordinating an experience to more general reasons, eliminating what makes it matter, what makes it *this* experience. Whitehead's schema gives the impression of coming out of nowhere, for it is the applications that prove efficacious for thought. They disarm it in a nonpolemical manner, separating each concept from its power to contradict.

Whitehead turned to metaphysics in order to make value the essence of the universe. His metaphysics has for its task to problematize each particular mode of valorization, be it that of a specific organism or of a specialized form of knowledge. Yet it cannot take the fact of holding together (the success constituted by enduring) for a sort of supreme value. That gesture would validate the stubbornness of the professional who refuses to let himself be troubled by what escapes his specialization. It would silence brooding that wanders without leading anywhere. Metaphysics cannot be a generalized ontology, if ontology has to do with the different manners of enduring proper to the beings that populate our realities. It is the task of metaphysics to ensure that nothing remains self-evident. And so it must also problematize the power that logic gives to contradiction, because contradiction demands stability of the definitions it puts forth.

What Whitehead calls *res verae*, or true things, are thus a metaphysical creation. The *res verae* or actual occasions of Whiteheadian metaphysics do not aim to give us access to a truth transcending cultures and specialized forms of knowledge. They are true in that they deploy the question of existence in a way that situates knowledge. They are not objects of knowledge, and referring to them will not enable knowledge of what is in question. In a manner of speaking, they constitute the elemental stitches of all that is in the making, or in the knitting. Each stitch puts into play that very particular feeling arising where continuity becomes problematic. It is not a question of rupture, but of a perpetual return of the question "how?" How to prolong, stitch by stitch? How to inherit? How to appropriate what is given? How to make it matter? "How?" is a question of "in what manner?" Didier Debaise calls Whitehead's metaphysical gesture "universal mannerism," undoing the opposition between subject and object.[12]

Obviously, the above questions tend to imply a subject. But, in this case, instead of a knowing subject, it is a subject whose existence is at stake. The manner in which it answers, in this way and not another, will *decide* who it is. Thus Whitehead problematizes what has, since the advent of modernity, been the site of brutal confrontation: decision versus causality. Either we are free to decide or it seems we are products of nature, subjected to causality. Only if we are free to decide, Kant thus stressed, are we moral beings, subject to judgment, responsible for our acts. This double abstraction (nature, responsibility) gives common sense a good deal to brood on, for it invades everything, giving rise to multifarious issues. To problematize is not to deny these issues, some of which actually matter. But it demands, above all else, to not make these issues into sites of confrontation between big contradictory categories.

After Kant, it seems, the notion of cause becomes such a category. Its role is to ensure head-on contradiction with everything associated with moral subjectivity. But it can function thus only if it gives a well-determined meaning to the notion of effect. This requires the possibility of defining "how a cause causes," defining a cause in isolation. This is never the case for Whiteheadian *res verae*. This is why they are also called actual occasions. If there be subjective decision, what must be decided or determined is how each cause will cause, *in this occasion and for this occasion*. Decision does not contradict causality. Decision concerns what will be cause for it, and how. In this sense, concludes Whitehead, "an actual occasion is *causa sui*."[13] An actual occasion entails appropriation of what becomes "its" causes. Causes take on concreteness. Their way of causing, or their role in the decision, is not definable in advance, in isolation. Causes no longer explain effects. They get their power to explain through the manner in which they participate in the occasional decision. The occasional decision gives its positive concrete determination to the subject, this subject. It also gives its full determination to what will have "caused" the subject to be what it is, feeling what it inherits the way it does. "The feelings are what they are in order that their subject may be what it is."[14] This is why Whitehead speaks of the process generating the decision as a "process of concrescence," or the coming into existence of an actual, individual entity, concrete and irreducible to any abstract explanation.

Now we understand at least one aspect of what Whitehead says we require to understand: "Each immediately present existence requires its past," not *the* past but *its* past. Existence prolongs the past in its own manner; this past becomes its past, the cause of *this* decision. And we can also understand that each existence "requires its future as an essential factor of its own existence." Whitehead's thinking of causality calls upon something the notion of cause was supposed to exclude: the end constituted by the coming into existence of the actual entity. Yet, like causality, finality is separated from all explanatory claims that would confer on the end the power of reducing efficient causes to simple means. Inspired by situations in which a plan, project, or goal preexists action, such a power introduces the hypothesis of an end susceptible of being conceived in abstraction from the action that achieves it. Here, biology is the battleground par excellence. For, in biology, the power of causes (whereby everything is explained by biochemical processes allegedly ruled by a blind efficient causality) is in conflict with the power of ends (whereby each process is supposed to play a role or to have a function in the service of the totality). What Whitehead names the doctrine of evolution has the misfortune of finding itself taken as responsible for fitting the two modes of explanation together.[15]

To characterize the final dimension of the process of concrescence, Whitehead uses the term "aim."[16] What the subject aims at is nothing other than its own coming into concrete existence, and it obtains existence by the same process through which causes obtain their determination. This means that causes themselves aim at the coming into existence of the subject they will be causes for. By the same token, neither the subject nor the ensemble of "data" the subject will turn into its causes can be thought independently of one another, as would be the case if the subject was thought in isolation, envisaging the swarm of data it inherits like so many scattered pieces of a puzzle.

Whitehead invents the term "prehension,"[17] highlighting the connotation of grasping also found in terms like "apprehension" and "comprehension." "Prehension" designates the subjective grasping of data in the strong sense of data asserting themselves as coming from elsewhere, attesting to an elsewhere foreign to the subject. It is up to this subject to make these data its own, to turn them into an integral part of what will be its world. As early as *Science and the Modern World*, Whitehead writes that the subject entails "prehensive unity" here of what is over there. As the demands of the metaphysical point of view unfolded, Whitehead encountered a dangerous abstraction that he needed to resist. The danger lay in considering data as such, separating them from the grasping that turns them into prehensions. Such an abstraction would encourage the (Kantian) image of a subject unifying, in a unilateral manner, what it has appropriated as its data.[18] Thus, in *Process and Reality*, prehensions, like the data they are the grasping of, may well be initially a nonunified multitude, a crowd-like many. Yet this crowd is not in the least indifferent, or alien as are their data. The unification it will undergo is not foreign to it, not imposed on it. As soon as there is prehension, there is involvement in a subjective process, along which the subject is in play as much as prehensions. The aim that prehensions will progressively realize "animates" them. Prehensions aim at their subject. Consequently: "These subjective ways of feeling are not merely receptive of the data as alien facts; they clothe the dry bones with the flesh of a real being, emotional, purposive, appreciative."[19]

As subjective appropriations of data, prehensions thus participate in the process through which the subject is determined, or determines itself. But the process of determination must be called problematic, bringing into existence an enigma to which no one possesses the solution, and which nonetheless insists that there be a solution. This is why Whitehead placed this solution under the aegis of a decision, which affirms the value of "thus and not otherwise." The solution obtained thus involves a process of exclusion. Certain prehensions will be "negative," not denied, but eliminated. This means that a totalizing point of view is impossible. Actual occasions are not reducible to perspectives on a same universe. Each occasion is a fact, the *creation* of a perspective, a value affirmed in a partial manner. The universe has no other unity than that conferred on it by the multiplicity of occasions, each unifying it in its own manner.

Thus Whitehead may write that "the unity of the universe requires its multiplicity." Without effective, discordant multiplicity, no particular importance could be attached to each unification such as it is achieved. If each unification could be situated or derived from an impartial totality, unification would be at once partial (*partiel*) and impartial. The "not otherwise" that Whitehead adds to "thus" must remain irreducible if the universe is to escape the impoverished static perfection of Grand Totality.

No genuine decision, however, leaves things unscathed. Exclusion leaves marks, its contribution to the process of determination. Exclusion imparts an emotional tenor to the decision, an insisting sense of what might have been but is not.[20] That there had to be exclusion is part of the achievement. It is what distinguishes this achievement from the happy imbecility of "so that's how it is." The emotional tenor of decision ensures that imagination will not be reduced to a supplement to decision. It will not be reduced to a fiction superimposed on the world that would merely be what it is. It affirms an irreducible dimension to the fabrication of what is. Humans, with their irrational dreams, truly are children of the universe. Faced with the irrevocable partiality of the noncancellable "thus it will have been decided," the importance they give to unrealized alternatives affirms that what is irrevocable will never have the last word. This does not separate us from a truer, fully immediate relationship to reality. It attests that the fact cannot be separated from the possible.

Still, the possible truly is excluded by what has been achieved. The achievement arising as the end of the subjective process of concrescence is both finite ("this") and "finished." It is finished because everything has been determined, which means that the process of subjective unification is finished. It thus also marks the end of the subject. Whitehead says that the subject perishes. What has been achieved is now an object among many for other prehensions. The achievement becomes a datum among the multitude of other data. Prehended, it will become a part of other processes, available for a future in which it will be a cause for something other than itself, "beyond its own actuality of realization."

Whitehead sometimes accentuates how the present is concerned with the future, how achievement itself aims at its beyond. Are "accomplished facts" the heritage of a past that the present must decide? Or are they making themselves part of the past *for* a future that will replay their limitations? "The creativity of the world is the throbbing emotion. Of the past hurling itself into a new transcendent fact. It is the flying dart, of which Lucretius speaks, hurled beyond the bounds of the world."[21]

This throbbing emotion may harbor within it what has been excluded, beyond the limits of the finite achievement, the "it will have been thus and not otherwise." The solemnity of the world as we experience it demands that "infinitude requires the finite," always this finite "thus" whose value is obtained at the cost of exclusion. It also demands that the fact refuses to be deprived of its relevance to potentialities beyond its own actuality of realization.

I will stop here. To continue in this vein would take us down a path that would defy common sense, so to speak. Whitehead-Socrates would never have expounded on metaphysics to the citizens of Athens. He would have prolonged their brooding in a mode likely to activate their resistance to the predatory categories that make experience bifurcate. The philosopher must try to weld common sense and imagination instead of giving lectures on the procedures of welding.

Whiteheadian Societies

Whitehead asked his system to make a metaphysician of him, to render him capable of discerning what coherence demands of us and obliges us to. Coherence is not the vision of a coherent world. It is the possibility of thinking that which we know while our modes of abstraction pull it apart. For me, Whitehead's system is a machine to make us think, but never in general, always in *this* situation, in the manner of the moral advice of Leibniz, *dic cur hic?* Thinking doesn't happen in a void. We do not extract ourselves from a situation to figure out how we think about it. Or, to be precise, even if we do step back to examine a situation, we do so in terms of the new situation that makes our examination matter. The metaphysics of actual occasions cannot be expected to provide answers to reflexive questions if they entail establishing grounds for the possibility of experience. Nor does Whiteheadian metaphysics provide answers to questions of truth, if this means truth as an object of contemplative enjoyment. Still less does Whiteheadian metaphysics provide answers to questions of communication if they are about the possibility of an understanding over and above the interests and convictions that separate us. Whitehead avoids the three traps of philosophy according to Deleuze: reflection, contemplation, communication.[22] Yet he escapes without anyone for an instant entertaining the possibility to figure him as the "Christ of philosophers," as Deleuze presents Spinoza. Whitehead is a mathematician, practitioner of the art of problems. For him, the problem must take on the power to situate the one who poses it.

When common sense places what makes it brood in the hands of the philosopher, this philosopher will not respond with a lengthy dissertation on metaphysics. The metaphysical concepts of Whitehead are precious for the philosopher because they provide her with tools. They allow her to escape the unescapable dilemmas and overblown false problems created by philosophy. Concepts are tools akin to the catalysts chemists use to produce reactions and compounds that would be otherwise impossible. So it is with Whiteheadian metaphysics. When the philosopher seeks to activate the welding of commonsense and imagination, Whiteheadian metaphysics offer the means to steer clear of the abstract obviousness that is part of the baggage of philosophy. That which claims obviousness stands in the way of the experience of sheer disclosure, which is the benchmark for philosophical success.

The same goes for the idea that existence must lend itself to being apprehended in a pure manner unsullied by what makes us wonder, as if it was a premise gifted with the power to assert its consequences regardless of the situation. Claims are made in modern philosophy for mere existence to be recognized as a condition for all else. Typically, mere existence is the "I" instituted by modern philosophers as the rightful proprietor of the experience it enjoys. The world in which we experience thinking—the "what" and "how" of our thinking, which particularizes all thought—must pledge allegiance to this "I." Their claims must be turned into mere correlatives of the "I." The radical impossibility of thinking reality for itself is carefully orchestrated and dramatized to silence what we nonetheless know.

> Mere existence has never entered into the consciousness of man, except as the remote terminus of an abstraction in thought. Descartes' "Cogito, ergo sum" is wrongly translated, "I think, therefore I am." It is never bare thought or bare existence that we are aware of. I find myself as essentially a unity of emotions, enjoyments, hopes, fears, regrets, valuations of alternatives, decisions—all of them subjective reactions to the environment as active in my nature. My unity—which is Descartes' "I am"—is my process of shaping this welter of material into a consistent pattern of feelings. The individual enjoyment is what I am in my role of a natural activity, as I shape the activities of the environment into a new creation, which is myself at this moment; and yet, as being myself, it is a continuation of the antecedent world.[23]

Has the *cogito* indeed been poorly translated? According to the *Thesaurus Linguae latinae*, *cogitare* comes either from *co-agitare*, "to continuously mix ideas together," or from *cogere* ("to gather together"): "The mind brings together many elements in a whole in order to be able to make a choice."[24] In brief, the etymology takes us back to an experience of actively struggling with a multiplicity instead of accepting the double abstraction of a bare "I think" in opposition to an equally bare extensive substance. Cartesian reflection affirms the double self-sufficiency of *res extensa* and *cogitans*, which do not mutually require each other. For Whitehead, this means Cartesian reflection is incoherent. The price for the proud affirmation of this incoherence is the omission of the individual enjoyment that comes with any process of appropriation to affirm that the subject of enjoyment is indissoluble from what it enjoys. But neither philosophy nor common sense have to ratify this operation. Neither need rest content to leave enjoyment to specialized forms of knowledge like psychoanalysis and neurochemistry.

Metaphysics is surely what activates Whitehead's characterization of the *cogito*. His account plainly draws on the multitude of prehensions, or feelings. It draws on the subjective unification of prehensions and the appropriative experience of enjoyment of the subject, which becomes a "self" feeling "its" world. But this characterization is not metaphysical. It is about human experience as ongoing appropriation. As such, this characterization is liable to dethrone the Cartesian "I" as rightful proprietor, and yet the metaphysical *res verae* do not have that power. They can neither dethrone nor confirm. Analogous to the stitches of what is being knitted, they remain mute with respect to the reality in which they participate. They remain mute with respect to how, once determined, they will participate in the knitting. They are metaphysical abstractions, what Whitehead would call "living" abstractions, because they make think, because they have been created to make think and feel in the modern world. The value of this creation depends, however, on operations of knitting and connecting. It depends on how it allows us to raise questions about the varied and multiple realities populating the world. It depends on how it activates ontological questions.

The ontological questions bring us back to the entities Whitehead called "organisms," the salvation of reality. In *Process and Reality*, however, he calls them "societies." The question now is: what does it mean for an actual occasion be part of a society?

An occasion does not persist in the sense of maintaining itself across time. It has its own temporality, that of becoming itself, which is essentially "atomic." Once it has become itself, the subject perishes as subject and what has been achieved becomes object, a datum available for subsequent prehensions. Each occasion is a block of becoming. A society, whose primary trait is to endure, but also to change, is composed of occasions that do not endure and do not change. Whitehead thus writes that societies are what enjoy adventures of change throughout time and space.

To offer another contrast, a society as such does not affirm a value or an aim. The metaphysical actual occasions are now what is giving meaning to these terms. A society, then, *does not aim at maintaining itself*. Herein lies an affirmation crucial to the adventurous empiricism of Whitehead. He discards those explanations that attribute a tendency toward self-conservation to social entities. To attribute them such a tendency would overburden what must remain *a fact*. Of a society, we may only say: the fact is that it has maintained itself *until now*. We should not imply that, so doing, it fulfills a "will" to persist. Correlatively, an environment is required for the maintenance of a society, yet how an environment *matters* for a society is also in the order of fact. It is a matter for observation or experimentation, not explanation or justification.

The relation between Whiteheadian metaphysics and its ontology passes through these contrasts. Using such terms as "value," "aim," and "decision" in reference to actual occasions means only not to submit them to any larger finality than becoming themselves, and especially to some whole that would claim them as its parts. A society, in contrast, is what maintains, across change, continuity of style or character, even if it be only for a short lapse of time. For actual occasions to belong to a society is to become in a way *conforming* to this continuity. Yet *nothing obligates them, nor even asks them to* respect this conformity. If they do so, it is rather because they are situated by what they inherit—that is, most of the other entities that provide them with data having already respected social conformity. *This is why societies matter.* For each occasion, they constitute a social environment. Societies situate the decision of each of their occasions, but they do not determine it. Each will appropriate what it has to inherit in its own manner, whether conforming or not. The continuity of a society is thus a fact with no general justification. A society is irreducibly individual, each with its own style, its own character.

The term "character" is interesting. I have already used it a good deal in verb form, "to characterize," because it makes no claims to define or explain, only to indicate ongoing conformity. The term is suited to moments when we say, "that's him to a T!," or conversely, when some behavior strikes us as unexpected, shocking, unusual, and we say: "Well, that's not like him at all! What is the matter with him?" Usually we ask fiction to maintain character. The writer's decision to make a character do something uncharacteristic must pay off through the ways in which protagonists of the story, intrigued, interpret and speculate about it. And if they feel that the writer has introduced something uncharacteristic on a whim, most readers feel frustrated.

For Whitehead, any description or explanation presupposes a "passage from experience of individual facts to the conception of character. Thence we proceed to the concept of the stability of character amidst the succession of facts."[25] Yet descriptions and explanations are always relative to a stability that nothing guarantees, to a "fact" that does not illustrate any generality. Explanation, here, always comes after, as a way of appreciating and commenting on the fact that "that" holds.

A society, writes Whitehead, changes but sustains a character across change.[26] Conversely, he adds, sustaining a character may allow us to characterize change. The great triumph of the sciences of Newtonian inspiration lay in their ability to identify how the societies they addressed maintain their conformity across changes of a spatiotemporal type. The law of gravitation offered one characterization of conformity. But the adventures of living beings are quite different. In the case of living beings, Whiteheadian societies prove useful because they intensify hesitations, countering the temptation to explain everything by reference to an invariant, for instance, to explain metamorphosis in terms of what remains the same. Consider a primate raised in a human environment, where everything has been done for him to learn sign language. Has the primate merely found an opportunity to reveal previously dormant capacities that can now be attributed to his species? Or must his handlers accept responsibility for participating in the creation of an unprecedented being whom it would be cruel to force to live in a world without humans to sign with? As for us, do we really know what we might be capable of given a different social environment?

Such hesitation, of course, concentrates on questions that often disregard that a primate or a human is not *a* society, but a vertiginous entangling of societies that collectively make up the environment for one another. The "character" that makes us hesitate ignores something that is of no interest to psychologists, ethologists, and writers either: the innumerable societies composing what we call a body. This body endures while we hesitate.

Whitehead characterizes a living body as that region of nature whose parts are centers of expression. Each part is intensely sensitive to others. Each part matters for the others. Thus, their individuality is irreducible. Parts cannot be explained by means of something other than themselves, by other parts for instance. Neither can they be explained by means of their role in the body taken as a totality. Integration (*faire corps*), then, is not a matter of each part collaborating in something that matters to all of them. Integration instead means participating, taking part. The parts are partners, so to speak. Parts relate to one another, deal with one another, matter for one another, and make a milieu for one another, *but each in its own way.*

Whitehead's hypothesis of centers of expression does not ask biologists to become metaphysicians. Something else is at stake in his hypothesis. As researchers' questions become more pertinent, what they characterize will look less like an organization in accordance with the human ideal of each part upholding and being held to its role in the service of the common good. The more the holding together of the body feeds the curiosity of biologists, the more biologists will make how it holds together intelligible, and the more what is required by the fact that it holds will surprise them. "The body is that portion of nature with which each moment of human experience intimately cooperates. There is an inflow and outflow of factors between the bodily actuality and the human experience, so that each shares in the existence of the other. The human body provides our closest experience of the interplay of actualities in nature."[27] Whitehead here makes a distinction between human experience (and, I might add, chimpanzees, dogs, and all other animals he refers to as superior), on the one hand, and bodily actuality as well as other actualities in nature, on the other. His distinction does not mark a return to the binarism of body and soul. It is pragmatic. Human experience, too, is "socialized." Still, in certain circumstances at least, our experience seems to require an address to "us," either to call us to order or to open a new perspective for us (sheer disclosure). Behaviorism utterly failed because it believed it could overlook this requirement. In the name of science, it managed only to accumulate facts extorted by apparatuses that were at times nothing short of torturous. In fact, some living beings demand to be treated as being "at the origin" of their ways of engaging with and answering to their world. This fact does not mean that we must define them (and ourselves) as subjects of experience in the metaphysical sense. Instead it asks us to learn how to wonder at modes of composition, intra-action, assemblage, which require this mode of socialization. What we have immediate experience of, however, is a flux. Whitehead sometimes calls it a stream, a clear reference to William James, but this stream of experience that integrates affects derived from the body has the particularity of giving importance to its own past, which experience "continues."

In *Process and Reality,* Whitehead dubs this type of continuity "living persons."[28] The term "person" must not be misunderstood. In the generic sense, personal order designates for Whitehead a single line of inheritance sustaining a character through the special genetic relation among its composing actual entities. This way of inheriting permits assigning a character to entities maintaining themselves through time.[29] The continuity of a person thus characterizes beings to which natural sciences attribute the power of self-conservation "by themselves," as well as the power to explain the rest. An atom is a person, for instance. But, in the case of a living person, it is no longer self-conservation that must be problematized, that can no longer be defined as self-evident, but the unceasingly recreated continuity of an ever-changing flux. "Between the bodily occasion and human experience, fluxes of factors enter and exit," and they do so in both directions. Certain words may kill. Means cultivated from the dawn of time—bodily techniques, dance and meditation, so-called psychotropic substances—transform how we understand ourselves.

Metamorphoses matter to us because personal living continuity matters to us.

The immediate experience of having a body should not become the battleground for abstractions that dismember it. Nor should it repeat the mind–body problem beloved of certain Anglo-Saxon philosophers who elevate to the status of a staggering enigma the impossibility of coherently rearticulating what their own operations have wrecked. With societies, it is a matter of learning to feel astonished by cooperation, by the multiplicity of modes of existential participation to which this experience testifies. But this does not confer on our experience any sort of foundational status. With Whitehead, nothing ever grounds anything.

And this is where Whitehead anticipates the objection that might well ring forth: "Our doctrine seems to have destroyed the very basis for rationality."[30] Indeed, how can reasoning have the least generality if we cannot trust in the self-identity of what he calls character? If it cannot be isolated from change? Do mathematics and logic not uphold the fact that numbers, for instance, remain what they are across the operations that mobilize them, and that individuals designated by logical variables are indifferent to the implications that concern them? (The Socrates who is mortal, since he is a man, remains Socrates.)

For Whitehead, such objections reveal limitations to the pertinence of mathematics and logic. Common sense, however, is not in the least threatened. Common sense finds nothing scandalous about the mutual necessity of change (process) and individual (character). It can readily agree that there is neither process nor individuals in general: "The form of process derives its character from the individuals involved, the characters of the individuals can be understood only in terms of the process in which they are implicated."[31] It is, in fact, the very art of much fiction to permit the exploration of this entanglement. Fictions allow the reader to discover how the character of each individual will be affected when put to the test by a change, which is also individually characterized. If the author is successful in rendering it, individual characters will be affected in a mode that agrees with what they have never ceased to be.

Whitehead's ontology connects a society with the maintenance of character through change yet does not permit defining this character independently of change, or defining change independently of what it affects. In fact, only rationalism will protest that we are destroying its foundation if societies are diverse and resistant to all generality in their relations of entanglement and interdependence. The problem of rationalism lies in the possibility of definition, while Whiteheadian ontology is connected to the possibility of practices that allow for apprehending and learning. Whether we begin with change or with character, we can learn. Change can allow us to characterize a society, and the character of that society can permit us to characterize the change affecting it. In this way, we may connect apprenticeship to the question of potential: how much change is a society capable of without being undone?

This question brings us back to the connection between Whitehead's ontology and his metaphysics. The maintenance of character does not stem from a power of self-maintaining. It stems from a multitude of occasional decisions, as each one affirms a "thus" that might have been otherwise. If the science of Whiteheadian societies is considered as sociology, his sociology affirms possibilities that are quite familiar to us in the context of human societies. Take, for instance, an army that previously acted as one falling apart in the heat of battle, and then it is each man for himself. Or, a social situation deemed normal that collapses: citizens who previously agreed on the necessity for public order may respond to the lure of adventure and explore the joys and troubles arising from new modes of partnership. Or again, outrage may inhibit any change that is felt as endangering social order. In these instances and others, explanation always follows from observation, for the explanation relies on what has suddenly been lacking. Still, even though we cannot explain societies, the character they maintain, or how they cease to hold, this fact does not situate us in an unintelligible reality. We proceed by detecting resemblances and distinctions amid diversity in that mode of abstraction called "analogy." This is why the inhabitants of Athens could answer Socrates's requests for definitions only through a constellation of concrete situations in which what they recognize as just and courageous takes on meaning.

If Socrates had encouraged them to discuss their analogies among themselves, they might have doubted certain ones and have begun to perceive them otherwise. Nonetheless, the outcome would not be the delineation of a good definition. Instead, it would be the situations of concern that would be enriched by other versions, that would have their imagination aroused about other modes of characterization. The invoked situation would cease to be an illustration and gain the capacity to make them think. Analogies are to be discussed. Certain ones may prove more pertinent or less, while others, deemed bizarre or out of place, may meet with derision. For Whitehead, attention to analogies, regardless of their pedigree, is the first and last word for what we call intelligibility. It allows us to make observations and inferences, to anticipate and nurture the contrasts that interest us and the divergences that intrigue us.

Rationalism itself proves as unable to transcend knowledge by analogy as was the famous Baron von Münchhausen to pull himself out of the quicksand by pulling on his hair. But that does not imply skepticism: if we cannot access the truth of things, we can increase our penetration, writes Whitehead, which is what the citizens of Athens might well have done if given the chance. "The procedure of rationalism is the discussion of analogy."[32] That discussion may lead analogies back to make-believe, cliché, or mere metaphor, but whatever is abandoned is not abandoned in favor of a so-called literal meaning. The analogies that are accepted will be acknowledged for connecting with interesting problems, for directing attention toward pertinent resemblances, and for proving their worth under difficult circumstances. That any analogy is problematic is not at all a weakness for Whitehead. Thought is an art of the problem.

Evidently, Whiteheadian metaphysics is itself a montage of analogies whose particularity is to undo the privileges associated with some of them and to dissociate them from their power of inclusion and exclusion. Every occasion, even the occasion belonging to the kind of person we name an electron, entails enjoyment, decision, animation by a subjective aim. Substituting society for organism is undoubtedly connected to the risk Whitehead perceived, that of organism as an analogy giving to the endurance of the whole the privilege of assigning a role to its parts. The explicitly analogical character of Whitehead's metaphysical concepts is consistent with the mission he bestows on philosophy: to weld common sense, which is under attack from those who deem it ignorant and impose requirements for abstract definition, and imagination, which enjoys the experimental use of analogies and the exploration of contrasting ways to characterize a situation.

Heirs to Whitehead?

It is now possible to state in another way the problem posed by the Galilean-Newtonian scientific lineage, which has so successfully welded positive scientific description and mathematics. Mathematics are now everywhere in modern sciences. For better and for worse, the multiplicity of their uses is a sign of the multiplicity of analogical modes operating in sciences. Yet, for physicists, everything begins with the mathematical definition of acceleration, and they have inherited from this success the deep-rooted conviction that what they call law points to an understanding that transcends analogy. The accelerated motion of a ball along an inclined plane proposed by Galileo initiated the science that was christened as "rational" mechanics, an appellation that celebrates the alignment of definition and intelligibility.[33] Falling as it did where Galileo had calculated that it "should" fall, the ball, or more precisely its motion, ratifies a hypothesis that is not about simple prediction. The motion of the ball testifies for the full and complete equality between a cause and its effect, the cause capable of identifying and measuring the effect, and inversely.[34] In other words, the motion provides its "reasons," which are full and complete; it dictates the manner in which it must be understood, it explains itself through its performance. Nothing calls for further penetration.

From the time of Laplace's demon through today, this event has transmuted into a distinct inability to discern between metaphysical statements and certain statements made by physicists. Kant wanted to domesticate this indiscernibility by leaving everything we can positively know to the imperium of a kind of generalized physics, while refusing physics any access to reality in itself. He would give space to analogy in *The Critique of the Faculty of Judgment,* but only in the mode of "as if." Humans are allowed to admire, to judge in terms of aims and values, as much with respect to art as to nature, but without confusing such judgments with the causal explanation that makes of them "only analogies." Thus, the status of secondary qualities from ancient philosophy is acknowledged, but it is also circumscribed, domesticated, rendered incapable of interfering with our objective definitions. The bifurcation of nature is thus stabilized. The confrontation is programmed between bearers of facts who explain by disenchanting (in the mode of "it is only . . .") and the defenders of the ineffable (inspiration, genius, etc.).

To inherit Whitehead, then, would be to succeed in thinking all knowledge as analogical without nostalgia for something beyond, and to recenter the question of rationality around *discussion* of such and such *always situated* analogy. Is it relevant? What does it make noteworthy? What does it omit or neglect? Does it initiate a relationship or authorize indifference? I will return to this manner of becoming heir to Whitehead by taking up cases Whitehead did not envisage. This is not only because such cases take part in more contemporary forms of knowledge, but also because they will help me to take up the relay, to become heir to Whitehead in a mode that activates the imagination a bit otherwise. I wish first to point out why I felt compelled to do so.

Previously I stressed how Whitehead felt himself as belonging to modern civilization, whose decline he diagnosed and whose adventure he sought to reactivate. Today we are living through the collapse of that civilization and, as if shipwrecked, are trying to determine what we should put in our lifeboats, what might be of use, what may prove to be of value in a future "beyond modernity." It is a difficult moment, and divergences abound, even to the point of conflict. But sorting through the remains of modern civilization is also a fascinating task of practical deessentialization that goes to the very heart of what is reputedly essential, asking us to distinguish between what can grow only in modern soil and what might learn to live in other soils. The question thus arises: is the science we call modern "essentially" modern, or can it become an ingredient for adventures other than those we have known of it? "Science can find no individual enjoyment in nature: Science can find no aim in nature: Science can find no creativity in nature; it finds mere rules of succession. These negations are true of natural science. They are inherent in its methodology. . . . Such science only deals with half the evidence provided by human experience. It divides the seamless coat—or, to change the metaphor into a happier form, it examines the coat, which is superficial, and neglects the body, which is fundamental."[35]

This assertion was made in 1935 in the course of two lectures in which Whitehead contrasts "nature lifeless" to "nature alive." The science he takes on in the lecture "Nature Lifeless" is the physics of his era, showing how it has culminated in the elimination of what it owed to direct observation: physics has rid itself of "enduring bits of matter" endowed with movement characterized by speed, in favor of "group-agitations" that "extend their character to the environment."[36] He notes that, now, "nature discloses to the scientific scrutiny merely activities and process," and Whitehead is always leery of anything said to be "merely," and asks of these activities that arise and fade away, "activity for what, producing what, activity involving what?"[37] The nature of physicists is dedicated to answering "nothing" to these questions. Physics must deny what its methodology asks it to omit, because the nature it addresses claims self-sufficiency and the capacity to explain its own functioning. In effect, this is the case for all sciences derived from Galileo and Newton: the equality of cause and effect makes it possible to assert that nothing has been omitted, the causality staged is indeed the whole truth of what is described, and the rule of succession exhausts all that may be known.[38] But Whitehead here subjects the natural sciences as a whole to judgment. And to characterize what they all omit, his privileged example is the experience of having a body.

The body belongs to nature; it is an object of biomedical science that asks it to explain its own functioning. But what is omitted is my body's experience of self-enjoyment as a living being. This enjoyment necessarily implies a certain immediate individual experience that is "a complex process of appropriating into a unity of existence the many data presented as relevant by the physical processes of nature."[39] The body is privileged for Whitehead because nothing of what concerns it is truly self-sufficient. In fact, as we all know, when our body is in question, the proof science requires of nature always entails a trial as well—heart aching while waiting for the diagnosis an analysis has permitted.

The argument seems incontestable in this respect, yet it is difficult to ignore its abruptness. What are those "physical processes of nature" that allow biomedical sciences and abstract physics to be put in the same basket? Everything happens as if Whitehead, impatient, suddenly took his reader by the collar, and shaking her, asked her to listen to what the abstractions that inhabit her make her say and ask her to ignore: "Nature is full-blooded. Real facts are happening."[40] She must dare to feel the self-enjoyment of existence and dare to think of it as fundamental, despite the bloodless dance of categories that make abstractions of it. How impressive is the philosopher when he knows how to raise a ruckus, throwing caution to the wind to bring us back down to earth!

Let us nonetheless beware the trick of evil, the insistence on birth at the wrong season, a trick that is so often provoked by impatience. It is indeed possible that Whitehead, in his later years, lost patience, abandoned the idea that his colleagues would ever agree their science was likely to omit something in a manner other than provisional. It is possible that he had lost hope in the possibility, expressed in *Science and the Modern World* ten years previously, that the notion of organism could provide a transversal concept unifying the diversity of sciences around the explicitly partial question of endurance. In any case, for the plurality of analogies, he has substituted a hierarchized opposition: "physical nature," what the so-called natural sciences decipher in terms of regularities and laws of succession, is superficial, while the individual self-enjoying, appropriative experiences that make up "living nature" are fundamental.

To be sure, Whitehead does not say the superficial is false. Science is neither a dream nor a falsification. Still, science needs to acknowledge that what it finds is but one factor of reality, calling for an experience that could be called more visceral.[41] "The doctrine that I am maintaining is that neither physical nature nor life can be understood unless we fuse them together as essential factors in the composition of 'really real' things whose interconnections and individual characters constitute the universe."[42]

But who is this "we" who will bring about this fusion? The trick of evil would come into play if scientists understood this fusion to be the business of philosophers. And it must be recognized that Whitehead is vulnerable here, because in the pages following the declaration of this doctrine, he successfully upholds a rather admirable indetermination between what our immediate experience of having a body affirms and the metaphysical construction proposed in *Process and Reality*, but without naming it metaphysical as such. The body in this sense becomes the ontological example par excellence of what metaphysics proposes, a "fusion" between what seem fated to contradiction: the privileged causes of physics and the aims and values that living beings require. This could be seen as a fine exercise in vulgarization. Yet we must add that, far from nurturing the broodings of common sense, our author proceeds directly to the implementation of an analogy whose goal is to persuade. Whitehead thus risks making his listeners (and his readers) the worried witnesses of a nasty duel between two authorities, science and philosophy.

The opposition between the nature of scientists and the experience of having a body is not a path I will follow. On this heading, I prefer to evoke Audre Lorde's famous warning: "The master's tools will never dismantle the master's house."[43] This statement, offered by an African-American writer, would, for instance, call into question the meaning of an amnesiac reconciliation between the children of masters and the children of slaves who were the property of those masters, under the auspices of critical rationality ("all is well that ends well"). Here I permit myself to hear in Lorde's warning not a verdict, but a proposition, which forces me to think as a philosopher, and in particular to raise the question of the task Whitehead entrusted to philosophy. Vigilance toward "our" modes of abstraction can mean conserving them as tools, and putting them to good use like experienced workers, which means not giving them the power to impose themselves, not giving the hammer, for instance, the power of seeing in everything a nail to be hammered. At the same time, the proposition of Lorde reminds us that, even if put to good use, a tool commits us, commits us thus and not otherwise. She asks us to answer for our choice of tools, forbidding us to assert that civilization "can be understood only by the civilized." The master's house was also the house in which or around which slaves worked, among them Lorde's ancestors, who had their own understanding of our civilization. She refuses to forget that.

To be sure, Whitehead never envisaged dismantling the house of the master. Nonetheless, if the task of philosophy in Whitehead's sense is not to transcend the civilization to which it belongs, then it is incumbent on him not to ratify the terms in which this civilization has been thinking itself. Yet the opposition he dramatizes takes us right back to Newton, Hume, and Kant, the three thinkers who set up camp on the solid ground of modernity. That ground was not only established epistemologically. It was established through the process of eradicating ways of inhabiting and living with the land that could never conceive of defining it in terms of private exclusive property. The process began in Europe and colonization extended it throughout the world. The right to property is the first tool of the master, which justified the ravages due to the alliance between modern states and capitalism, blessed by reason and progress. Whitehead indeed posed a problem that Hume wished to avoid and that Kant resolved by eliminating: how to situate "our" perceptions in a world that functions for its own account. But the direct analogy Whitehead proposes between his metaphysics and the experience of having a body places the emphasis on the experience of the individual who meditates on the meaning of his rights of property: we know that we are in the world, but we also know that this experience of the world is *ours*; we know that this experience prolongs a past, but also that this past is *its*; we know our emotions derive from bodily activities, but also that it is *we* who feel them.

Whitehead may well try to complicate what we know, to deny that the owner has rights transcending facts. Yet, to propose direct access to metaphysical comprehension, he relies on what may seem undeniable to every modern individual. In doing so, he situates us in the same rarified landscape as the thinkers whom he wishes to resist. Even to discuss them, he employs tools or operative analogies that today we must refuse to forget are those of the master.

To put it another way, Whitehead indeed inverted the epistemology of Hume and revitalized the older metaphysics that Hume had claimed to eradicate by requiring abstract sensible impressions to prevail over speculation. But an operation of inversion always implies an element of conservation. Forcing the point a bit, we might say that the metaphysics of actual occasions here finds itself domesticated, set firmly in place through the evidence proposed by a reflexive "civilized" consciousness, prone to the temptation of dualism, inhabiting "its" body as the master inhabits his house, without having to worry about what ensures the permanence of its functioning, and watching over external nature through the windows that filter out what is effective and transform it into inoffensive sensible impressions.

Of course, Whitehead took care to note: "When we consider the question with microscopic accuracy, there is no definite boundary to determine where the body begins and external nature ends."[44] But these sorts of scruples do not divert him from his priority: to use Hume's tools in order to dismantle his doctrine. He does not dramatize the possibility that a brick just smashed the windows of the master's house!

Yes, in my peaceful den, I can share Whitehead's surprise over the necessity for me to think that I am in this room, and at the same time, this room, in one way or another, is "in me," an element of my present experience, of what I am now. But this astonishment is conveyed primarily through a sensoriality dominated by the paradigm of vision. The point of view shifts back and forth between my point of view on my environment and the point of view of the virtual enunciator for whom I am part of nature insofar as I occupy a position in this environment.

Nevertheless, as we shall see, Whitehead's metaphysics does not privilege a depopulated world. Neither does it privilege a sensoriality that would let abstraction triumph. On the contrary, his decision to make the demand of coherence prevail forced him to think outside well-trod pathways. When it came to matters of doctrine, however, he mobilized metaphysics to intervene in a conflict for which the masters' analogies had already supplied the tools. For me to relay Whitehead, then, instead of my experience in a world as peaceful as my den, what I need to understand is the difficulty that I have, the "civilized" me, with letting myself be effectively affected, touched, concerned by *this* world that our modes of appropriation have ravaged and are still in the process of destroying.

To relay something is not to turn one's back on it, however. It means inheriting something otherwise, in this situation that is ours today, in which we no longer know very well what it means to inhabit the modern world. Of one thing I am sure: we must not act as if we could wash our hands of the modern world, renouncing that heritage and seeking to recover the innocence we associate with worlds deemed "primitive," to be modernized. This is why the contrast Whitehead omits, between science and sciences, is important for me, in addition to the task Whitehead associated with philosophy at the time of writing *Science and the Modern World*: not to construct a doctrine, but to cultivate vigilance toward our modes of abstraction.

Trying to give scientists a taste for vigilance with respect to their modes of abstraction is perhaps more necessary today than ever. But it is equally indispensable for us to avoid activating the trick of evil: in an increasingly troubled season, we need not enjoin scientists to recognize that they are letting the essential slip away. For, they are threatened on two fronts. First, they are in the service of an innovation-hungry economy that ridicules a distinction between the fundamental and the superficial. Second, an aggressive skepticism goes after anyone who sounds the alarm to warn us about the consequences of the history of predatory innovations we call development.

When it comes to deessentializing the sciences or, as Bruno Latour puts it, bringing them down to earth, do we not need first to activate the imagination of scientists in a mode that distances them from such abstractions as Science and the Scientific Spirit? Do we not need to try to detach them from the model of triumphant rationality proposed to them by this particular science, the legacy of Newtonian physics that literally ignores the earth and its histories as lacking laws worthy of the name? This lineage tolerates other sciences only with the hidden caveat that their findings remain merely relative to human interests. They are not to be compared with the kind of knowledge physicists claim, the one hypothetical extraterrestrials would also obtain if only they take an active interest in the functioning of things.

If we return a moment to the metaphor of the coat (or the superficial) that Whitehead proposed, may we not interest ourselves in the art of the tailor, who cannot ignore the body of his client, even if he takes that body into account only in relation to his own aims that clothing falls well, that it suits the animated morphology of the body? In contrast, is not the manner in which Whitehead characterizes science, similar to that in which we might characterize the *prêt-à-porter*, or worse, the imposition of a single model on all bodies, indifferent to its suitability?

In fact, we might go so far as to say that the fusion between nature lifeless and nature alive, whose necessity Whitehead affirmed, has taken place in a certain way but not at all in the mode he envisaged. I am thinking here of the cooperation between James Lovelock, originally a chemist, and Lynn Margulis, a biologist, who together invented Gaia, an Earth whose inhabitable character involves and requires the living beings that populate it. If the laws of physical chemistry alone had been in command, Earth would have been as uninhabitable as its neighbors, Mars and Venus.

Lovelock received a great deal of criticism for the analogy he proposed between Gaia and a living organism, because, contrary to living beings, his Gaia is nourished only by sunlight, does not reproduce, has no congeners and still less prey, and is not the outcome of any process of selection. Gaia has the traits of an organism in Whitehead's sense, however. Its habitable character is not derivable from general laws: it is a fact, a continuous individual achievement. It is inexplicable in the sense that nothing more general than it explains it, that it is sui generis, but it is intelligible, explaining itself as "engaged in its own immediate self-realization."[45]

If something like fusion between nature lifeless and nature alive has indeed occurred, for scientists it has happened in the sense of creating problems of a new style, connecting with new analogies. For instance, the endurance of Gaia does not constitute the finality pursued by the living beings who participate in its continued existence. The *habitable* character of the earth implies and results from the inherently anonymous interdependence of a multitude of activities, physicochemical as well as living, none of which has a particular relation with a common interest that would contribute to explaining them. In other words, to construe their participation in the maintenance of Gaia in terms of "enrolement" is not a suitable analogy. Enrolement, the assignment of a role, is the biologist's daily bread, but it requires a process of selection. The behavior of each component of a living being has been selected for its role in the capacity for survival of the lineage to which this being belongs. But there is no population of Gaias, and thus no competition for survival. It is here that Margulis intervenes, showing not only that Gaia is habitable, but also that it has been rendered *viable* for a wild diversity of manners of being alive by the invention of multiple relationships of interdependence, symbiotic or mutualist, through which living beings require one another to obtain what will permit each of them to live. The analogy of *partnership* not only fits the body as Whitehead characterized it; it designates what, for Margulis, permits an understanding of the manner in which living beings have participated in the existence of Gaia. It is through such partnerships that the components arising from general processes described by chemistry were captured and metamorphosed into resources for living beings, and Gaia became *fertile*.

Today the distinction between habitability and viability is critical. Gaia may well remain habitable for a multitude of inventive microorganisms that will permit her in turn to endure, regardless of the frenzy of "development," as we have christened it. On the other hand, when activists in the street declare, "we are not defending nature, we are nature defending itself," they signal the threat of destruction that weighs on innumerable ways of living on earth in mutual becoming, each thanks to others and at the risk of others. They speak of the destruction of the ongoing fabric of partnership all the way down that we call nature.

From the point of view of what Whitehead called science, times have changed: physicists may well continue to measure the value of their theories by reference to an imaginary community of hypothetical extraterrestrials arriving at the same conclusions, but they can no longer make us forget this Earth, with its entangled histories in which they too participate, even if their laws make abstractions of them. Some figures with scientific credentials have given us a taste for a science that makes us understand entanglement. Margulis recast the history of living things in terms of the invention of symbiotic partnerships. Rachel Carson sounded the alarm about the cascading consequences of apparently rational interventions, the massive use of DDT (Dichlorodiphenyltrichloroethane) to eradicate mosquitos deemed responsible for the propagation of malaria. Biologists and ecologists today discover entanglements at all levels. Causes are everywhere, yet none of them is capable of defining its effects, "all other things being equal." In fact, on Earth, things are never equal, and only rarely can causes be defined independently of what they participate in. Even the notion of system now feels too reassuring, for a system sets up relationships that are well-defined and stable in relation to the functioning they explain. The exploration of earthly entanglements requires critical care in the discussion of analogies, of what they make matter and what they take the risk of omitting. In other words, it requires vigilance toward our modes of abstraction.

To inherit Whitehead today is thus not to inherit a doctrine that counters scientific doctrines. His own metaphysics does not entail any particular doctrine—or a particular ontology. The temptation of metaphysical doctrines and of ontologies is that they encourage modes of characterization allowing what matters to those who propose them to prevail. Indeed, Whitehead himself did as much when he privileged the experience of having a body and, in a more general way, proposed a characterization of societies highlighting their relationship to an "external" environment, which, even by his own admission, constituted a gross simplification. If Whiteheadian societies do in fact have a mode of existence, it would have to be called problematic. Societies today call for those who strive to characterize them to consider how they belong to a fabric of interdependencies. They call for researchers to see themselves not as situated within an environment, but as belonging to this fabric, and thus as liable to destroy it. They call for researchers to learn how to participate in the weaving of the fabric and accept that they are themselves woven in it. When activists declare that they are nature defending itself, they not only resist the bifurcation of nature but also commit themselves to learning what a culture of care and attentiveness requires.

Some scientists today actively participate in calling into question our modes of abstraction. A new generation of biologists has confirmed and continued to enrich Margulis's proposition that the invention of symbiotic relations is the signature of life on earth. As what was omitted begins to count, new analogies flourish. Implied in this fact is the possibility of sciences themselves participating in what I have called their deessentialization. It might then be possible to be heir to Whitehead in his protest against the explanatory claims of the sciences without providing sustenance for the trick of evil.

{ 4 }

What Can a Society Do?

How to avoid activating the trick of evil when speaking of sciences? How to avoid freezing researchers in a defensive attitude? Such an attitude is understandable, given that researchers feel themselves exposed and subjected to imperatives that put them directly in the service of ensuring economic growth, while confronting a public that traditional institutions are no longer capable of disciplining, which seems to confirm their worst prejudices. When calling into question the bastion that goes by the name of Science, how not to run into blockage and inhibition, expressed as a kind of social panic, "let's hold on, not give an inch, or chaos will ensue"?

Thinking through the Milieu

The situation is not a war, however, in which each must choose their side. On the one hand, so many researchers would like to work on questions pertinent to these times of ecological and social collapse. On the other hand, the catastrophe-orientated vision according to which "people" would now think that facts are mere constructions is rather an exaggeration, even if many people are more likely now to question dismissals based on "there's no proof," which might previously have silenced them. We find ourselves in the type of situation I have associated with the possibility of a "diplomatic" intervention: belligerent parties are caught up in a logic that seems to make war inevitable, and yet, if the diplomats make them feel that it might be possible to avoid it, they might opt instead to suspend the reasons pushing them toward war, allowing the diplomats to give peace a chance.[1]

As the diplomat knows, diplomatic intervention is possible because the reasons for war are "social" in Whitehead's sense. Logic and its "therefores" are not enough to deduce the necessity of conflict. The situation that gives logic its power could be characterized otherwise. In the case of Science, the reasons for conflict come at least in part from how scientific institutions encourage scientists to adopt a manner of self-presentation fated to pit them against the public, whose perspective becomes characterized as meddling in what it cannot understand. It is worth noting that this manner of self-presentation carefully omits the tensions arising among researchers over the imperative for science to serve economic growth. The public, such as it is characterized, would only too easily follow the example of those who take facts seriously only when facts are in keeping with their interests. Or else it would be only too happy to reject any scientific claim as corrupted.

Instead of making a frontal assault on the reasons mobilizing scientists, which would activate the trick of evil, the diplomat strives to inflect them. Diplomacy is a pragmatics: it is an art of effects, an art based on the fact that no reason, as such, has the power to determine what it requires. Considered in the terms proposed by Whitehead, we could say that diplomacy counts on the fact that how a society defines what makes it hold together remains an open question, which only the trick of evil turns into an inflexible self-presentation. In the terms proposed by Gilles Deleuze and Félix Guattari in *What is Philosophy?*, we could say that the diplomat has to think the reasons justifying the war "through the milieu," exploring how these reasons imply and capture a propitious milieu that nurtures evidence for them.

In *An Inquiry into the Modes of Existence*, Bruno Latour turns to the figure of the diplomat in a mode of thought experiment whose stake is to dramatize a discrepancy the brutal effects of which he has amply experienced. Through his fieldwork, he acquired an appreciation for the intelligence of practitioners of science who themselves were thoroughly happy for the opportunity to share their ambitions and concerns with an inquirer who, while posing questions that surely surprised them at times, displayed genuine interest in what they were doing and what mattered for them. During the great warlike mobilization against skeptical relativism, however, Latour found himself the butt of reprisals on the part of the same practitioners (or others like them) who suddenly offered the public a dogmatic caricature of what they were doing. Refusing to recant, to recognize that "there is an objective reality, which science merely discovers," he was denounced as a supporter of irrationalism whose ascendency constituted a threat to civilization.

The diplomatic intervention Latour conceived is situated by this experience: he undertook not only to understand better how certain formulations had proven capable of inspiring such anxiety among practitioners but also to address the striking discrepancy between their practices and the ways in which they demand to have them recognized—or it means war.

The diplomatic apparatus, as Latour imagines it, entails an attempt to activate possibilities for peace by putting in place a very particular milieu, an agora, the public place par excellence where Greek citizens listened to orators disputing among themselves. The agora he imagines would gather protagonists concerned by the possibility of formulating ways of speaking well about forms of knowledge locked in rivalry today, each trying to disqualify the other or coming to an agreement only in order to do away with yet other forms of knowledge. The diplomat in the agora must successfully "speak well of something to those concerned by that thing—in front of everyone, before a plenary assembly."[2]

Instead of inquiry, diplomacy involves an encounter with specialists in public, precisely where they are used to engaging in belligerent behavior. Scientists of diverse provenance will be there, as well as other practitioners, legal experts, theologians, and doctors, to name a few of those who belong to institutions that tend to hold public self-presentation at arm's length. And since diplomacy is an apparatus, it is all about seeing how it can function. How is the assembly to be composed? How will roles be distributed? How can it activate what the trick of evil inhibits?

A first protagonist is the one who has taken the initiative to hold the assembly, taking up the diplomatic challenge of speaking well before the others. Next are the "belligerents," those whose respective ways of presenting their practices would mean war against each other if they did not avoid explicit confrontation. Without the initiator who has invited them and who professes to speak to them of what concerns them, they would not have gathered, because their habit is instead to "speak badly" of one another but in situations where they enjoy impunity, without consequences, in cold-war situations. Of course, those who take pride in their arrogant judgments, the hardened professionals, have refused the invitation. And finally comes what I have been referring to as the public, those who have grown accustomed to being taken as witnesses, or hostages, or empty-headed straw men. The belligerents agree only on this latter point. But the public today is troubled. Let us have the members of this public be chosen by drawing lots among volunteers in order avoid reconstituting cliques. The presence of the public, with its way of composing a "milieu" for the other protagonists, will be crucial. It will, in effect, give meaning to an injunction Latour inherited from Whitehead: "The goal always is not to shock common sense."[3]

This injunction, distinctive to the agora, also dramatizes the consequences of what I have previously called the defeat of common sense. The belligerents count on a public that is patient, unaccustomed to feeling shocked, even if today it is feeling uneasy and disoriented, and even if furious impatience has overtaken some of them, as we are all too aware.

Let us imagine this assembly. Each of those whom I call belligerent parties, when confronted with the challenge of a diplomatic proposition concerning it, will have to consider implications and consequences. The wager of Latour, diplomat, is that each practitioner is accustomed to presenting her practice to people deemed incompetent and incapable of understanding and is primarily concerned with maintaining an apparatus of "territorial defense," instead of explaining what they are doing in practical terms, elucidating how it matters to them. Even if a practitioner does not possess the blind arrogance of the true professional, he feels obligated to speak in terms of belonging to an institution, which is mobilized to defend instituted boundaries, and which strives to command the respect of outsiders, whose ignorance constitutes a threat. If he becomes caught up in giving an account that highlights the specificity of his practice, he knows that his peers might very well disavow him: "You got played, you've weakened us."

Instead of a faithful empirical description of the landscape of practices, the "categories" in play in the agora correspond to the diplomatic problem posed there, which does not concern all the practitioners; many practitioners are prey to territorial struggles, habituated to living on the sidelines, in a subaltern position of formal or effective dependence. Consider, for instance, the relationship between those who take care of zoo animals and are acquainted with them and the researchers who come to observe these animals "objectively," or the relationship between farmers and the technicians of an ever-more-rationalized and industrialized agriculture. Caretakers or farmers know only too well the experience of being characterized pejoratively, yet they are not in the habit of presenting themselves as aggrieved or dissatisfied, and this is why they are not part of the belligerents; they are thus in the public.

Let us continue as the situation becomes increasingly complicated. One of the singularities of Latour's agora is that actually concerned practitioners, practitioners belonging to "predatory" institutions, are gathered together in the agora, and each has very good reasons to listen to the proposition the diplomat addresses to another, and to the reply. The exchange may concern him because he knows that this other may present his practice in a mode that often implicates him as well, and generally in an unflattering manner, explicitly or by default. Typically, science, then, is neither politics (a tone of disgust), nor religion (disgust again), nor a human convention like law (irony). Science is the one true (vibrato) source of technical innovation, and whoever likens it to mere technique, relative to the interests it serves, is the enemy (exclamation). Usually, this sort of knocking down of a straw man remains without consequences—think of the false courtesy that makes for the boring sterility of so-called interdisciplinary encounters. But in the agora, in the presence of a public that is as interested in their silence as in their interventions, each of the parties is concerned. Each must pay attention to the manner in which the diplomatic proposition addressed to another and the answer made to it implicate or enlist them. Tolerance no longer works; mutual ignorance is not an option. The staging requires a kind of choreography in which the protagonists, accustomed to bumping into each other or stepping on each other's toes, would learn the art of encounters, attunements, and distances.

Latour's agora, moreover, requires the presence of the public in a mode analogous to that of the Athenian citizens, whose civic duty was to attend the spectacle of passions staged through tragedies and brood with the chorus commenting on the events. This public is not a direct stakeholder in the problem the agora dramatizes. But it listens, cogitates, and appreciates. It is vigilant. And if this public is shocked, it will be because of a pejorative judgment, of concern precisely because insulting to it, or else because, while it is brooding, it is shocked by authoritarian stupidity, a hasty shortcut, a dogmatic dismissal, or a blatant inconsistency on the part of those who claim to know. It is because the public feels empowered to evaluate and wonder that it introduces a constraint. The potential impatience of the public forces the belligerent parties to seek ways of speaking well, of not relying on predatory judgments to characterize what they are doing, what it demands, and how it situates them. The presence of the public even incites them to listen to one another, to let themselves be affected by the ways in which each of them tries to respond to the challenge.

Can the parties reach a point where each is capable of presenting itself, of speaking well about what matters for it, without any need to define itself against others? Latour's story does not say. But if they are able to do so, it is not because they have successfully come to respect common sense as one might respect traffic regulations. As I have emphasized, when Whitehead describes the welding of imagination and common sense as "a restraint on specialists,"[4] he does not mean someone slamming their foot on the brake pedal when running into the radar measuring their speed. At stake is the channeling of their imagination.

Deleuze and Guattari ask us to think through the milieu. Here we need in particular to think through the rarified frictionless milieu required by the institution called Science. Science makes it the right and even the duty of the specialist to resist the temptation of being affected by the objections of others or their perplexity, and so this rarified milieu produces predators and prey, those who judge and disqualify and those who are victims, whose reasons are dismantled derided or marginalized. The agora truly is a testing ground, because the way in which specialists themselves represent their practices is inseparable from this rarified milieu, inseparable from their adherence to the necessity of distancing themselves from uncertainties that might entrap them in common-sense questions that they take for mere opinion. Distance is all the more required because getting interested in matters of common concern might make them lose time or lead them to hesitate in an unproductive mode that does not advance knowledge.[5]

In Latour's agora apparatus, not shocking common sense thus means speaking before an attentive public, a public that is demanding and alert, likely to take offense if it is treated as a herd in search of its master, a public that hypothesizes an "amateur" milieu whose participants are capable not only of taking interest and evaluating but also of objecting. Without recognizing it, modern institutions desperately lack such a milieu today.[6] Previously I alluded to Greek tragedies, which staged the unfolding of fatal passions, indifferent to their consequences. These tragedies were considered to be essential for the political culture of citizens, for citizens had to acquire some knowledge of these passions, to taste them, in order to avoid subjugation to them. The speculative character of Latour's agora underscores the absence of such apparatuses in the modern world, apparatuses nourishing a political culture capable of distinguishing between what we call democracy and the art of leading a herd.

The modern world inspired Whitehead to call for cultivating vigilance toward our modes of abstraction. The apparatus of the agora gives new meaning to that call for vigilance. Instead of a task for philosophy, maintaining such vigilance now calls for a change in milieu. Recall that Whiteheadian societies depend on the patience of the milieu they affect. Here this means making a wager on a milieu empowered to feel its impatience when faced with hegemonic claims that shock it. What might practitioners prove capable of if their institutions of affiliation did not spare them the challenge of this impatience, if they did not teach them to remain narrowly focused on the interests of their practice to the exclusion of all else? Conforming to this injunction, specialized practitioners may be highly skilled and innovative in their domain yet ignorant, naïve, and arrogant about the rest, what Whitehead called "professionals." Latour's wager implies that such conformity is not essential to modern practices. Exposed to a demanding (yet not accusatory) milieu, practitioners might become capable of exploring other ways of characterizing themselves and their practices.

Today, the agora as Bruno Latour imagines it might appear to belong to a dead past. Diplomats could have intervened at the time when the science wars were waged, when the public still paid attention to specialists. But is not such a public on the verge of extinction? Panic now overcomes the herd previously held in check by promises of progress and economic growth as well as the inexorable advance of knowledge that placed the planet at the service of humans. The agora designates a possibility that today seems eclipsed by the urgent need to mobilize in defense of specialists against the hatred of facts.

To be sure, because specialists now are attacked when their facts prove bothersome, encouraging them to abandon their institutionally inculcated defensive habits brings to mind Jean de la Fontaine's fable about the schoolmaster who reprimands a child who is drowning. Specialists might well take up the retort from the fable: "Hey, friend, pull me out. You can yell at me after." In our case, however, the "after," which is supposed to arrive when specialists are finally respected again, is also in vital need of them becoming capable of respect for others as well, abandoning their dream of a herd that is calm, trusting, and governable once more. Mobilizing in an emergency, deriding the possibility of peace in the name of the harsh reality of priorities, is a dangerous advisor.

I hope that I made clear that thinking through the milieu is not at all about conforming to what the milieu makes probable or improbable. It is instead about resisting explanations that normalize, that anesthetize the imagination, and that give free reign to the trick of evil. It is especially about refusing to retroactively validate the claim that critical attention with respect to the authority of scientific facts amounts to a threatening rise of irrationality. Thinking through the milieu does not mean denouncing institutions as such, but characterizing modern institutions through the rarified milieu required by this authority, institutions promoting the myth of "isolated facts." In this context, it thus means bearing in mind that a situation demanding such rarefaction is dangerous and profoundly unhealthy, but without separating the situation from what Whitehead called its potentialities. Instead, thinking through the milieu is about the kind of option that William James characterized as genuine, alive and insistent, of importance to us, and above all, forced. For James, when an option is genuine, we are forced to opt, to make a wager, because "there is no standing place outside the alternative."[7] To abstain is to take sides against what might be. By dramatizing the possibility that specialists may betray the institutions that make professionals of them, Latour's experiment with the agora opts for the possibility that what we call our civilization may have a future. Latour thus pursues a path Whitehead would not have rejected.

I now head down another path, however, different from Latour's yet complementary to it. Both may be taken, each in its manner, as applying the famous Spinoza dictum to Whitehead's societies: "We do not know what a body can do." I will return to Latour, but in his exploration of what it takes to "face Gaia." The path I will now follow starts from Whitehead's impatient opposition between "nature lifeless" and "nature alive." To learn to speak well of sciences is also to learn to speak well of the way in which they address what is habitually called "nature." To deal with this habit diplomatically means to avoid rushing into oppositions that are likely to push scientists to adopt a belligerent stance. It will mean wondering about so-called "natural societies" and the way scientists explore what those societies can do. How do scientists learn to make sense of the reckless plurality of the social ways to make an environment matter, a plurality that makes up our worlds?

Finding More

Diplomacy is an art of words. The diplomat does not fear words, but understands their danger. She will not attempt to proscribe certain words, but will focus on detecting uses that make them dangerous.

Let us return to the bifurcation of nature. The question no longer concerns the bifurcation between objective facts and values or other subjective add-ons, at least not for now. The question bears on this ambiguous and controversial word, "nature." Some point out that nature no longer exists in the sense of being free of any trace of human activity. Plastic waste permeates even the ocean, and in the sky, the increasing amounts of debris threaten the security of space flights. Others primarily fear the connotations of the adjective "natural," understood to mean legitimate, normal, or authentic. Others yet reject the opposition between nature and culture, as much for its connection to human exceptionalism (a world in which the others may be globally qualified as "nonhumans") as for its theoretical imposition of a great divide between those who know how to distinguish between nature and culture and those who mix them up. And all the while, physicists in ultratechnologized laboratories continue to define themselves as being in search of the laws of nature.

The idea that nature should be "pure" or "wild" in order for her to "exist" obviously attributes far too much to human activity. Even bonsai, meticulously "civilized" by the Japanese, are not products of human activity. Neither are GMO monocultures, any more than are shepherd dogs. Humans have to go through what a tree, a plant, and a dog require to live if they wish to produce bonsai, GMO, and shepherds. The limit case may be industrial animal farms, where animals are kept "alive"—that is, prevented from dying, with doses of medications. For physicists in contrast, purity, authenticity, and abuse are of no concern: the earth moves, and the adversaries of Galileo can do nothing about it. This is nature for physicists, something indifferent to our ideas about it, which their laws strive to characterize.

It is utterly impossible to salvage any sort of "good definition" from this jumble of meanings and issues. What is needed instead is to create meanings that suit the needs of those for whom these meanings are of concern. As early as *The Concept of Nature,* Whitehead sought to create a concept of nature assuming "as an axiom that science is not a fairy tale."[8] Today we would say a science that is not a mere social, linguistic, or cultural construction. The nature scientists refer to must be able to provide them a *grasp* of it. Nature must not be reduced to a mute reality, thus relegating responsibility for what is attributed to it to the famous subject of knowledge alone. Neither must nature be something only scientists have access to. *The grasp nature provides them is a matter of concern for many others as well.* We thus need to ask what type of grasp the sciences require. "We are instinctively willing to believe that by due attention, more can be found in nature than that which is observed at first sight. But we will not be content with less."[9]

The "we" here is indeterminate. Indeed, it could equally well comprise nonhuman animals, paleolithic humans, or scientists. The attention of an animal on alert, the rabbit's ears directed toward the least sound, attests to the relation between perception and the possibility of finding more. Does the rabbit make a conscious decision to prick up its ears? No more than we do, no doubt, when we react to a sudden noise. In any case, that which we call consciousness intervenes only afterward in such cases, unless we have made a conscious decision to remain impassive (but even then micromovements would betray us to those with the means to decipher them). If nature is, as Whitehead proposes, articulated with the notion of a grasp enabling us to find more, it is not an object of conscious representation, but of pragmatic concern. The rabbit's business here is a matter of survival. Its way of paying attention implies the vital importance for it to distinguish between what is only a noise and the real approach of what it must flee.

Of course, what we are now calling nature entails different stakes for the fleeing prey, the approaching predator, and the observing experimenter. All of them, unanimously but each in its manner, reject the hypothesis of an inconsistent nature, a kaleidoscopic fabric of dreams that is made and unmade, transforming each time the way of paying attention changes. If they speak the Whiteheadian jargon, they also make clear that the stakes articulated around nature will be social, implicating occasions as they participate in societies instead of actual occasions as metaphysical abstractions. Thinking in terms of "finding more" forces us to confront ontological questions that hinge on the question of knowing what the different societies making up the world in which we live can do.

The expression "due attention" is equally indeterminate. Due to what, and why? The experimenter's attention has little to do with the rabbit's, or with the ethologist learning to decipher animal behavior, and yet, in the three cases, attention is due to find more, or so it is hoped. At the same time, there exists a number of other ways of paying attention, implying other modes of grasp, other relations, and other stakes than finding more. In those cases, sticking to Whitehead's proposition, I will no longer speak of "nature." This term will be reserved for that which is implied in the possibility of finding: a practical differentiation between the ones who seek to find and what they aim at. To find *more*, it is imperative for what is aimed at to remain itself, to remain that with respect to which finding occurs. Although to find is not the only way to learn, it is the best suited to the scientists' insistence on defining their successful outcomes in terms of accessing reality in itself, independently of their own anticipative interpretation. For all the exaggeration of their claim, we can understand it in terms of what a successful grasp requires for the one attempting this grasp. The alpinist who seeks a grasp with which to pull herself upward needs it to bear her weight. For the scientist, the ambition of finding more requires that the one about which there is finding be indifferent regarding what this finding entails. *Eppur si muove*, whatever the theological problems it raises, claims Galileo.[10]

Although the proposition may seem bizarre, it merits attention for the effects and problems it raises. In other words, its efficacy arises from pragmatics: it addresses a practice from the question of what it requires without turning this requirement into a defining condition. Indifference is relative to what the scientist aims at finding, it does not define "nature" in general.

Whitehead's proposition excludes assertions whereby "nature does not exist" or "nature is inaccessible to knowledge," which insult scientists, spurring them to adopt a belligerent stance. Indeed, how could we learn the attention due to find more of something that does not exist? How can something providing a grasp be called inaccessible? Nonetheless, that to which scientists' attention gains access does not constitute an authority for others whose attention implies different stakes. The concept of nature allows scientists to resist those who would reduce what they obtain to mere constructions. But it does not permit them to forget that they are situated by what they ask of nature.

Any grasp allowing a connection engages both parties concerned. If a realistic value can be associated with scientific practices, it comes from their mode of engagement, and if we can associate nature with the possibility of finding more, it is to the extent that that to which we pay attention agrees to the offer of engagement in the required mode. It cannot be a matter of seduction, nor of common interest, nor of obedience, nor of a unilateral imposition. Thus, we again find scientific objectivity associated with the theme of the event of a successful outcome. Sciences do not presuppose the bifurcation of nature, but when there is a successful outcome, a bifurcation passes between those who are actively responsible for the offer and the one who was empowered to turn it down and has accepted it. The story of a science is a story of "making societies," of recruiting, not conquering, what has lent itself to its grasp.

At the same time, when it is a question of the sciences we call experimental, the offered relation bears features of an enrolement that requires a maximum of dissymmetry between the two poles, between the one who acts, chooses, or develops and the one who must play its part without endorsing it as a role. The answer obtained must be identifiable with an impartial result: the being that answers must be indifferent to the stakes articulated around its answer, and thus indifferent to the role it takes on. The answer must be assimilable to a simple result of the experimental intervention.

The singularity of the experimental fact, then, lies in exacerbating the opposition between the one who asks questions and what answers. The experimental fact articulates ways of "making society" that authorize this opposition. The successful experimental outcome implies a passionate, intensely purposeful way of "making society" on the part of experimenters, since the success of an enrolement that meets the demands of proof really matters or has value only for them.[11] At the same time, this successful outcome is able to implicate only societies that let themselves be enrolled in an indifferent mode: for such societies, the laboratory is an environment like any other, where they just go on in their own way. The indifference required by an experimental setting cannot be obtained by simulating neutrality or cheating about what is at stake. It is an ontological requirement.

The term "successful outcome" is of diplomatic importance here. While it may satisfy experimenters, it separates them from those for whom proof is generally required, whatever the price. A successful outcome is not a right. The experimental finding demands a mode of attention that may not be the *due* attention in many other cases.

> An angry man, except when emotion has swamped other feelings, does not usually shake his fist at the universe in general. He makes a selection and knocks his neighbour down. Whereas a piece of rock impartially attracts the universe according to the law of gravitation. . . . It is true that the rock falls on one special patch of earth. This happens, because the universe in that neighbourhood is exemplifying one particular solution of a differential equation. The fist of the man is directed by emotion seeking a novel feature in the universe, namely, the collapse of his opponent.[12]

We can take the fall of the rock as a particular example of a general law established experimentally. But when the man punches his neighbor, the idea of reducing the punch to the impartial result of an ensemble of causes that scientists should actively identify is but the chimerical dream of neurophysiologists. Still, the man in anger is not a free and responsible subject. It is impossible to pass without transition from the laws of nature to the laws of humans that may judge him guilty. Whitehead construes the man's gesture as an example of purposeful animal behavior. A hungry tiger pouncing on its prey would have served him equally well as an example. What the two have in common is that their behavior cannot, unlike the movement of the stone, be characterized in abstraction from what they aim for. As regards behaviorists who feel it important to deny such aims and to reduce behaviors to an indifferent succession of actions and reactions, Whitehead remarks: "A consistent behaviourist cannot feel it important to refute my statements. He can only behave."[13]

The example of how experimental sciences "make society" is thus not generally pertinent in regard to animal behavior. The man's fist might well aim for the experimenter himself, if he found out that the latter is in fact responsible for the provocation that aroused his anger, that he was "manipulated" or "baited." To be sure, the capacity to be sensitive to decoys or lures is not the prerogative of all living beings, but it signals what could be called the end of experimental innocence. The animal, too, might well be capable of finding more, and notably with respect to the apparatus that is supposed to enroll it in the service of science. The meaning of science then mutates: the apparatus is now that to which the animal gives meaning, in one mode or another, yet always in a partial manner, for its own reasons.

Those who, "outside science," learn to know an animal, always *this* animal, may well be indifferent to the question of knowing whether or not what they learn responds to the values allowing science to affirm that it has found more with respect to nature. For ethologists, however, the question is critical. They would really like it, for instance, if what they call "habituation" allowed them to claim that animals have become indifferent to their presence and have resumed a "natural" behavior. The term "nature" becomes dangerous as soon as the dissymmetrical enrolement that allows for "finding more" becomes tied to the idea of nature having its own laws or natural behavior having its own rules. The same trap occurs with words such as "obeying," "resulting," or "being indifferent." "You see, you broke it," said Epictetus to the master who mistreated him. But the indifference of the stoic does not make him into an instrument that tests the limits to the resistance of a tibia. *Perinde ac cadaver*, or obeying in the manner of a corpse, was the vow of obedience taken by ascetic monks. The obedience of the monk who renounces his own will does not make him an inert body. Even Pasteur's and his successors' experimental success does not depend on the indifference of microorganisms to experimentally controlled variations in their milieu. On the contrary, success depends on the decidedly piecemeal, partial manner in which they evaluate this milieu. Pipette, petri dish, and living body matter little to the microorganisms that lend themselves to this type of cultivation. But food utterly matters. If they accept the offer, they will provide a determinate answer to the question: does this milieu permit them to grow and multiply? This is, together with that of poison, the question that articulates the relationship between the experimenter and his population of microorganisms.

The term "partial" is central to Whiteheadian ontology. Every society is partial, making its environment matter in its own mode, valorizing it in this way and not any other. The fall of a piece of rock is an exception; it is an example of impartial behavior. Where it falls depends only on its initial conditions, and as a consequence, it may be the instrument of a crime or an example of misfortune that had someone in the wrong place at the wrong time. Impartiality means that aim is foreign to a body that is subject only to the law of gravity (parachutists who take advantage of friction, or birds, gliders, and airplanes, are another matter entirely). It is only because the piece of rock, as it falls, is indifferent to its effect that cause and effect can be defined as equivalent.

Privileging ties between "impartial" movements and the notion of mathematical function is normal enough.[14] But it is a disaster for them to be accepted as the model of intelligibility, implying the illusion of self-sufficient science denounced by Whitehead. From the point of view of Whiteheadian ontology, any analogy privileging impartial movements or changes is undue, for it does not activate the attention suited to finding more. A prime example is the extreme partiality implied in efforts to attain impartiality.

Thus, in each case, vigilance is needed with respect to analogies and the modes of abstraction they promote. Through the question of due attention, the plurality of societies with which we pragmatically relate comes into question, as does the plurality of practices likely to find more, notably those that are about "learning from." Here we take leave of nature, but without forgetting the practices of "learning with," for I will address them later.

Learning, of course, is what scientists do all the time, yet when their success communicates with the verification of a theoretical representation, the necessity of learning is often set aside. Physicists like to repeat, with respect to chemistry, that chemistry learns while physics understands. Indeed, chemists "learn from" the dizzying variety of chemical compounds to characterize what their ingredients are capable of. They may need, for instance, to take into account the possibility that a compound relies on the presence of a catalyst: even the separation between a possible operation and an impossible operation is relative to the circumstances. Intelligibility, when it occurs, is obtained after the facts: chemists *render intelligible.* And the intelligibility that they obtain is always relative to trials, to the ever partial manner in which a chemical body is likely to make its environment matter, to affect it and to be affected by it. "What is gold?" asked the ancient chemists, and their answer was that gold is what resists all acids, with the exception, since the eighteenth century, of *aqua regia.* Other answers followed that multiplied the whys and wherefores, but always relative to new trials, to the development of new manners of affecting. Chemists render intelligible what they learn, and today they do so especially thanks to quantum theory, yet they do not apply this theory; they use it and tinker with it to interpret what they have learned from what chemical bodies prove capable of.

For Whitehead, all knowledge is relative to the thing as finite, to this and not to that, but what is finite is not for that reason an object of definition. Whitehead himself tells us that there is "infinitude" in the finite.[15] "We can never fully understand," he writes, but then adds, "we may increase our penetration."[16] Thus, as the means of paying attention is geared toward what chemical *agents*[17] can do and what they are sensitive to, the way toward a veritable ethology is paved that reveals the fine interdependence between "molecular individual" and its milieu. Heraclitus said there are gods even in the kitchen, and chemists might well agree, were they to think "in the laboratory." To render intelligible, then, is not to define; it is to complicate all definition, to learn to be astonished by what seemed explicable, to find more with respect to what enters into the characterization of a finite thing. They learn from the manner in which it detaches itself from an environment with infinite entanglements but cannot be isolated, extracted from this environment that it implicates within its own limit.

The notion of increasing penetration may also refute analogies that might otherwise seem persuasive. It was necessary to dissociate the sun from the idea of a power exerted on planets, for instance, to accept that the sun does not attract the earth without the earth attracting the sun as well, and with the same force: the difference between the star and its satellite depends solely on their respective masses. Similarly, when one pinpricks an inflated balloon and hears it deflate with a pop, the obvious analogy is that the air rushes outside, as if it had regained its freedom stifled by its confinement within the balloon. Since the nineteenth-century triumph of what we now call statistical mechanics, an increased penetration leads, however, to denial that the air "escapes." The behavior of a gas is understood as the result of a multitude of radically irregular individual motions of molecules colliding with each other. The air does not escape. Molecules are defined as indifferent to the way out offered. A molecule crosses the breach if its fundamentally erratic motion leads it to do so, and that's all.

A regular, reproducible behavior can thus result from sheer irregularity. What seems to obey mathematical laws may be understood not in terms of obedience, but as the manifestation of the indifference of an aggregate of societies to the individuality of these societies' adventures of change throughout time and space. What can be described mathematically are statistical "average values" that express the fact that, in such "crowd societies," individualities may be averaged out or smoothed over. They correspond to a simple noise with no measurable consequence. The notion of nature lifeless would then make sense where science need not take into account individual aims.

Are the erratic motions of the molecules really all, however? When those erratic, aimless motions create a vortex, when a tornado goes its own way, destroying all in its path, do such phenomena not call for physicists to learn more? Does it not seem that the erratic motions of molecules composing a tornado take on a kind of togetherness, as if a new "sociality" had emerged from the crowd of molecules?

If the emergence of a tornado were not happening empirically, the event would be deemed impossible. Still, for physicists, the fact of its occurrence does not give it the power to refute the erratic character of the molecular motions that constitute it. Researchers would unanimously reject any explication of the tornado implying that molecules suddenly feel each other mutually in such a way that their individuality starts to count so that they can move together, self-coordinating like a flock of birds or a school of fish, for instance. Nor would they be wrong: asserting that the motion of a molecule may suddenly cease to be erratic would kill the problem. The tornado requires physicists to find more on the subject of crowd sociality, such that they *learn from* the possibility of a tornado how to complicate the notion of crowd. The tornado's intelligibility should not require new roles assigned to individual molecules, but a problematization of the characterization of a crowd by statistical mechanics specialists.

In this case, to put it briefly, what is problematized is the pertinence of the notion of average value, the bridge built by statistical mechanics between the "law" the gases seem to obey and the crowd of molecules composing gases. The notion of average value implies that the overall behavior results from behaviors indifferent to one another. But what does this notion of indifference depend on for its validity? Specialists in statistical mechanics relate this notion to the possibility of dividing a system into microregions that should be "uncorrelated," which means that a local deviation relative to the average has no, or only negligible, repercussions on other regions. In contrast, the emergence of strong, long-range correlations marks the appearance of a form of social sensibility that does not imply agents being affected by one another in a new way. It implies agents in crowds behaving differently. Posed through the notion of correlation or of repercussion, the question now bears on the crowd as such: does what happens here make a difference there? The analogy of the crowd is enriched. Instead of being an answer, the crowd has become a problem: what can a crowd do?

This may lead to yet another manner of dramatizing the difference between the intelligible world of laws and the "finite things" that the sciences seek to *render intelligible,* from which they seek to learn. The tornado has the power to intrigue physicists (and frighten us), but it has not been offered a direct explanation through ad hoc capacities lent to molecules. Instead, what takes on meaning, what becomes pertinent, is the notion of *circumstance.* In what circumstances can a crowd become "sensitive to itself"? The explanatory power of laws is not supposed to depend on circumstances, but as soon as intriguing cases arise, circumstances intervene. When it is a matter of learning from, due attention bears on what is particular to this case.

When it is a matter of sciences, the gods that Heraclitus placed in the kitchen are well placed. The kitchen is a place where laws do not demand obedience, but participate in an art of composition to be negotiated and flavors to be obtained. When declining to explain technicalities, scientists often speak indeed of "kitchen" problems, and this often designates the operations through which they bend their theoretical representations to the exigencies of cases that are intriguing in relation to theory. These operations aim to render intelligible that which they address, implying a double, correlated transformation: there is the transformation of the scientist who, instead of applying his theory, questions it; and there is the transformation of that which scientists characterize, which must obtain, depending on the circumstances, the capacity to participate in the intrigue.

The fact that scientists speak of the "kitchen" also signals that these transformations, through which general laws are separated from their claim to authority, should not concern the consumers. The distance to be maintained from a public deemed unable to understand commits scientists not to give too much publicity to practices that tamper with laws to make them pertinent to facts. Such practices do not enshrine a knowable nature, but they give full importance to the question "what *here* is due attention?"

I do not aim to relay Whitehead by diminishing the grand success of the Galilean-Newtonian paradigm. I prefer to leave such success to its beautiful solitude and fragility. Its realization of the coincidence between intelligibility and submission is relative to a rarified milieu, purged to the utmost of what can never be entirely eliminated, *friction*. I prefer to think with cases in which intelligibility is gained, where tension suffuses the way in which scientists understand what they study. To be sure, such scientists never attribute aims or values to what intrigues them. Still, they put in place agents that raise questions about how they are to be characterized, what they are defined as capable of, what they are defined as sensitive to, through the always partial manner in which they act and interact according to circumstances.

Caring for Analogies

When Heraclitus's gods are in the kitchen of models instead of the scene of evidence, it is sometimes difficult to know whether a researcher speaking of an experiment speaks about an actual one or a computer simulation. In the kitchen of models, nature is not required to make science possible, to give authority to definitions. The model enables negotiation with agents that are put into action to see what they can do together. A model does not have general authority. The model translates a situation in such a way that everything is placed on the same plane: laws constraining action, if there be any; particular circumstances; and agents as they have been hypothetically characterized.

Scientists today study a nature that is populated with agents, and multiagent models are typically used to simulate intriguing behaviors. Each agent is characterized in terms of what it does, and in so doing, what it makes others do, or permits or prohibits them doing. The agent here is not a purely fictional construct. It is conceived in terms that allow its behavior to be understood as a hypothetical function of what are called laws of nature: it "behaves" without an aim or a capacity for self-determination. Nonetheless, the relation to the law has changed. The actions of the agent no longer demonstrate its submission, but matter from the point of view of their consequences on other agents. Always in a particular milieu, these actions put the law into action, so to speak. The agent actively deals with the milieu it "perceives" and "experiences."[18] In other words, rather than being considered subject to the laws of nature, agents are characterized as answering to a situation. Models thus put to work an understanding by analogy. What is more, depending on the case, they provide means of discussing whether an analogy is well-founded: if characterized in this way, can the staged agents render intelligible the emergence of the intriguing collective behavior in which the model supposes they participate?

Thus in some situations a car may be seen as the instrument for a dream of individual autonomy: going where I want, if I want, when I want. When motorists are stuck in a traffic jam, however, everyone stopped or inching forward, the will of each motorist, whatever it might have been, finds itself frustrated. The cars' motions now depend on one another. If the traffic jam has a cause, such as an accident, there is nothing intriguing about that. Sometimes, however, congestion happens without an external cause. The so-called multiagent model can show how, based on a threshold of density, a certain regime of stop-and-go traffic may arise, marked by transitions between the regime of fluid, laminar circulation in which cars are instruments of individual (human) will (that is, are driven fairly independently of one another, each with its own individual speed) and the regime where everything bottles up, in which individualities are flattened out and individual aims become insignificant. Cars are still driven by humans, of course, who, in rage or serenity, keep their individuality, but the question of the model is that of the transition between situations that allow some to drive as fast as possible without any overall consequence, even if others are forced to slow down, and situations in which the reasons for speeding up and slowing down no longer count and the dependency of each on the others prevails.

Multiagent models may serve to problematize the modes of abstraction suitable for characterizing the behavior of agents in situations of interdependence. Each agent is characterized in terms of what it makes others do in doing what it does, rather than in terms of the aim its action pursues or does not pursue. This is why such models tend to cross the boundary between phenomena we call social and phenomena we call natural. We can certainly denounce the perverse enjoyment of transgressing the difference that must be respected between conscious human agents and the others (that is, almost everything else). Modelers do not cultivate this enjoyment, however, except when it serves to shock common sense. What preoccupies them instead is to avoid endowing their agents with capacities that would explain in an immediately intelligible manner that which it is a matter of *rendering* intelligible.

The agents put into action by such models must not be confused with the "really real" things (*res verae*) of Whiteheadian metaphysics. Neither natural sciences nor the sciences of the mind have ever dealt with actual occasions. The actual occasion does not have the power to render intelligible, nor does it have power over anything whatsoever. It has but the power of becoming itself. In other words, that with respect to which we can eventually find more, what it is a matter of trying to render intelligible, only ever concerns societies. In analogy with a Whiteheadian society, agents staged by a model respond in their manner, in their own partial mode, to the solicitations of their milieu. They are defined by this partiality, determined by what matters to them, and how. But they do not define themselves or aim at their own definition. If there is an aim in play, it is that of modelers. When they define how an agent acts on and is acted on by others, modelers test whether the resulting collective behavior offers a reliable analogy with the kind of observed behavior they wish to render intelligible. Nonetheless, its assigned definition must not serve the ideal of an explanation that would lead to forgetting the intriguing character of the observation—that would be to find less.

Recall, for instance, how Whitehead characterized the living body in terms of centers of expression. If researchers characterized such centers by their role in the service of bodies, they would find less. But neither should they accept that the body may be reduced to an ensemble of processes ruled by general laws but that they are assembled, like a clockwork mechanism, in a way designed to ensure the functioning to be explained. Intrigue would be replaced by mystery. This is also the modelers' concern when they conceive their agents: to find more, they must avoid all trivial or circular explanations. The analogies they resort to for characterizing them must not lend their agents too much, but just enough.

In other words, when scientists protest, "but that would be anthropomorphism!," their reaction is not necessarily a fearful knee-jerk response. It does not necessarily mean that they are in the grip of the bifurcation of nature. In *Facing Gaia*, Latour speaks of the bifurcation of nature as fundamentally anthropocentric, since it leads to "deanimating certain protagonists called 'material' by depriving them of their activity and overanimating others called 'humans' by entrusting them with admirable capacities for action."[19] Scientists who protest about anthropomorphism are not necessarily rejecting models that involve a degree of "animation" of their nonhuman agents. What makes scientists protest and evoke anthropomorphism is the fear of "overanimation." But it may be that they accept for themselves this overanimation against which they protect what they seek to render intelligible. Why make anthropomorphism a sin? Why tolerate overanimated anthropocentrism?

As we have seen, overanimation is in evidence when the centers of expression evoked by Whitehead are characterized as animated by the aim of serving the body. But the political source of this analogy, a citizenry supposed to serve the well-being of the city, is also a typical case of rhetorical overanimation.[20] As soon as agents are attributed with powers of action or passion from which directly derive what is to be rendered intelligible, there is nothing to problematize: the answer kills the question. So many pseudosciences function in this manner. Consider, for instance, the "rational" economy, which provides agents dubbed "economic subjects" with perfect information and a limitless power of calculation, but devoid of memory and blindly egotistical in aim: this series of traits precisely allows them to be enrolled in the service of that abstract *prêt-à-porter* christened the market.

In *Facing Gaia*, Latour considers how to attribute animation to agents, speaking of transactions intervening in metamorphic zones for the renegotiation of both the competencies lent to agents and the researchers' mode of due attention.[21] In light of such transactions, so-called anthropomorphic analogies are no longer a sin, but a risk. Such transactions free scientific approaches from a methodology inherited from Newton, Hume, and Kant that, for Whitehead, condemned science to superficiality. But they do not agree with Whitehead's opposition between the superficial and the deep. They do not demand that a scientist be an explorer of depths, but rather an essayist, like the tailor who essays the fabric and the draping that are best suited to the body she must dress and who finds more to this body through her essays. If we enter the kitchen and cultivate an appreciation for transactions and the manner in which intelligibility is gained, it becomes possible to engage with scientists diplomatically, honoring their capacity to become (gradually) intrigued.

But, when the researcher-tailor deals with what living beings are capable of, she is no longer the only one raising questions. Aims abound, human and nonhuman, and transactions become difficult to manage. Whitehead contrasted the punch made by an angry man aiming to knock down his neighbor with the indifferent laws presiding over the fall of a piece of rock. When what we call anger is in play, analogies multiply. Anger overcomes someone, like a river overflows its banks. Anger explodes, like the outcome of a chain reaction: petty words, petty barbs, mounting irritation, leading to the blow-up. Such analogies require critical discussion, however. When engineers seek to channel a river, for instance, they can certainly say that the river will "take advantage" of the least flaw or the smallest oversight to thwart their plans. Nonetheless, whether dealing with a river, a chemical explosion, or a nuclear explosion, they treat these "powers of action," as Latour calls them, in a manner that implies a character implacable to the point of indifference toward what matters for us. Faced with the rising waters, prayer may help, but she who prays will not refuse the sandbags that might protect her house. In contrast, like the angry man's fist, the leaping tiger who takes advantage (in the proper sense of the term) of the inattention of his tamer introduces an aim with consequences. Trying to tame a river by enclosing it within concrete embankments and trying to tame a tiger are different professions, and the attention proper to finding more will not be the same for the engineer and the tamer: the latter knows that the tiger, too, is attentive, waiting for his moment.

What about living beings, like the tree or the oyster discussed previously, whose aim, it seemed to Whitehead, is almost exclusively the business of survival, indifferent to the consequences of this business for anyone but themselves? What happens when a scientist devotes to such living beings the attention that he thinks is due? How about the microorganisms Pasteur learned to cultivate in his laboratory? And what are we to make of the pigweed (or more politely, *Amaranthus palmeri*) that became resistant to the herbicide Roundup, flourishing in fields of genetically modified soybeans and turning into a nightmare for farmers? Obviously, we can say that pigweed did not aim to become a superinvasive weed; it remains ignorant of the human aims it thwarts, even though it really and truly took advantage of the selective pressure that the massive usage of Roundup put on it to innovate. The recklessness associated with industrial monoculture calls for words other than those we could use for the engineer or the tamer.

When it comes to beings whose aims they cannot ignore, biologists often lose the intelligence needed to care for analogies. The notion of intrigue no longer suffices, for the contrast between normal and remarkable does not come into play in the case of living beings.[22] Living beings *are* remarkable, and so it is often a matter of showing that they are "less" remarkable, and that a general explanation can or should normalize them, an explanation in the order of *prêt-à-porter*, which in this case means natural selection acting on genetic transmission from generation to generation. Usually, resistant pigweed would not be taken as the subject of action, but rather as the site of an "accidental" genetic event that Monsanto scientists and experts, following orders, claimed to be impossible, or more precisely, too improbable to be taken into consideration. The proliferation of resistant plants was an accident deemed arbitrary, but which had highly meaningful consequences due to the intensive selection through the use of Roundup.

Nevertheless, we may also say that the possibility of modifications, which may be called, borrowing Deleuze and Guattari's turn of phrase, "popularizable,"[23] has nothing accidental about it, but is actively implied by the business of survival of living populations (and in particular those in which the rate of potential multiplication is elevated: while pigweed cannot compete with bacteria of course, each plant still produces some twelve million seeds per year). The possibility of pigweed acquiring tolerance to Roundup thus is consistent with an adaptation its population is capable of when its milieu has become toxic. Science made in Monsanto neglected what the aim of survival makes the pigweed people capable of, just as the unbridled use of antibiotics neglected the intriguing capacity of bacteria populations to swiftly produce resistance against what poisons them.

"You have only to adapt!" Violence lurks in this injunction as soon as it makes of the milieu that to which conformity is demanded. But such violence first expresses contempt for what adaptation requires. When the river escapes the embankments to which its currents were supposed to conform, no one would dream of saying that it did not successfully adapt. A river does not adapt; in any circumstances, it does all it can do. In contrast, the indignant surprise of Monsanto scientists and experts that "it successfully adapted!" signals the major achievement of living societies, which Whitehead considers their very signature. Not to adapt to a milieu in which the massive usage of Roundup has become systematic is, for pigweed, synonymous with eradication, which is exactly what the use of herbicide aims to achieve. The resistant variety took advantage of the milieu ravaged by human intervention in order to proliferate. And it was able to do so only because it foiled the eradication aim, responding in a novel, *original* manner to what was supposed to be toxic for it.

For Whitehead, the possibility of an original response to what is socially given is what makes the difference between living and nonliving societies. And the possibility of this response requires, as we shall see, an irreducible reference to *res verae*. Nonliving societies lend themselves to an analogic characterization in the mode of "as if," to explanations reducing the individual fact to a simple result, abstracting it from its "relevance to potentialities beyond its own actuality of realization." In contrast, where there is life, the possible does not refer to speculations alone, or to human calculations alone. What we hold against so-called weeds is their capacity to transform the damage wrought by human intervention into opportunity.

With living beings, the importance of the possible is thus affirmed. To place the responsibility for adaptability on arbitrarily selected variations is to eliminate the real fact: adaptability requires agents capable of letting themselves be affected, at their risk and peril, by one aspect or another of their milieu that has till then been indifferent, and to attribute a new role to it. That is also to render irrelevant probability calculations that assume that what happens results from a closed ensemble of aspects of a situation retained as relevant.

Stuart Kauffman, a theorist of complex systems, has stressed a generic contrast between the evolution of living beings and the evolution of systems conforming to what Whitehead called a rule of succession. Evolution conforming to a rule of succession is conservative in that it conserves the definition of what this rule retains as relevant. It is this conservation that allows a model, be it deterministic or probabilistic, to be a tool for forecasting. In contrast, a model appropriate for evolutionary biology cannot be conservative. What the evolution of living beings forces us to think is an "expansive" dynamic in which the number of pertinent aspects of a milieu is not given but may gradually increase, in which the ways for a living being to make its milieu matter are likely to multiply and to become entangled.[24] In other words, the model requires the modeler to take interest neither in the given alone nor in actual relations between agents alone, but in that which the given can make possible, or "occasion," the opportunities it can offer to agents.

Kauffman thus affirms that histories of living beings require us to envisage a world in which the difference between actual and possible, between what is and what might be, is at stake for those beings for whom this difference matters: living beings. And we may characterize events like a becoming relevant, or the emergence of new ways for living beings to count with others and for others, as *original* in the sense of originating from agents for whom they matter.

> If I am right, if the biosphere is getting on with it, muddling along, exapting,[25] creating, and destroying ways of making a living, then there is a central need to tell stories. If we cannot have all the categories that may be of relevance finitely prestated ahead of time, how else should we talk about the emergence in the biosphere or in our history—a piece of the biosphere—of new relevant categories, new functionalities, new ways of making a living? These are the doings of autonomous agents.[26]

There is, to be sure, a great difference between the business of survival to which the oyster or pigweed is dedicated and the survival of humans, for whom it is about living for a "diversified worthwhile experience." Still, Kauffman's generic proposition—making a living—holds for both. Making a living, Kauffman writes, must last somewhat longer than the fleeting instant. This imperative, to have to make a living, recalls the difference Whitehead established between lifeless and living societies. The lifeless tornado is not self-sufficient; it must be fed by its milieu. But being fed is in the order of fact; it is not something it strives for. The analogy about a tornado aiming at "destroying everything on its passage" does not withstand critical discussion. In contrast, a living being actively "valorizes" its milieu, extracting from it what it needs to sustain its life. For a living society to make a living demands the destruction of others, and this is also true, maybe especially true, for living societies that are in need of a diversified worthwhile experience, at the risk of attributing to themselves the very experience of worth, of taking themselves as the source of all values. "Life is robbery," writes Whitehead.[27]

Is there robbery when a river, "valorized" by the concrete walls that enclose it, is deprived of its peaceful banks? The question is undoubtedly poorly posed. Because posed abstractly, the question is condemned to transforming the distinction between living and nonliving into an opposition, thus highlighting the indifference of the river to the embankments that would constrain it. But this indifference makes an abstraction of the fact that the river is not only a flow of water in the hydrodynamic sense. It also forms a milieu for a multitude of living beings, including human beings who valorize it, each in their manner: there are those who lay the concrete, and others for whom the river is a site for relaxing on the banks, observing, fishing, boating, or dumping trash, or a site of memory or encounter, a nuisance, and so forth. As the "river" entangles a multiplicity of modes of valorization, it is far from being indifferent to the "robbery" reducing it to a flow of water to be tamed. "Not everything is alive, but there is life everywhere," writes Leibniz.[28]

For his part, Whitehead declares, "the robber requires justification."[29] His writings have no need for a fable celebrating harmonious coadaptation, a form of equilibrium in which the robber would be justified by the good he conditions. What justifies the thief is nothing other than the generic contrast underscored by Kauffman, which Whitehead dubs "originality," the emergence, for better or worse, of ways of valorizing, of making matter, or of bringing into relation that are ever more partial and diversified. From this point of view, the act of unilateral valorization of the river as an indifferent flow and the bias toward eradicating "weeds" put into action by Monsanto's Roundup are so many manifestations of originality, their sad commonality being that they present themselves as justified and they proclaim their innocence when things go wrong. "That's not what we intended" is the refrain of those who define their intentions as justifications.

It is impossible to repeat the point too often: aim is not intentionality. Intention is one of many stories humans tend to tell themselves, and it is one that often proves impoverished in comparison with the many other stories that we need to learn to tell, as Kauffman reminds us. Still, however impoverished or remarkable a story may be, none of them is innocent, nor can any of them claim absolution. Instead, to say that life is robbery implies what Donna Haraway dubs "response-ability," a capacity to respond for and respond to, foremost to open ourselves to questions about the "sacrifices" that are justified by our intentions.

Thus, even when scientists may justify what they do in terms of finding more and showing the relevance of original ways of understanding, no such justification should be considered to absolve them. It is not merely a question of the means they must mobilize or the ways in which their knowledge will be mobilized. Obviously, when they deal with living beings, the ambition of finding more is not innocent when they practice animal experimentation. But learning from an animal what it is capable of, as a representative example of its species, is not objectionable in itself. It is definitely not a matter of pronouncing the guilt of the ambition for finding more. It may be, however, a matter demanding utter vigilance toward our modes of abstraction. Never should the ambition to find more exclude the legitimacy of other kinds of questions. The questions posed in order to find more have to render themselves able to make room for other questions that engage she who poses them otherwise. In such cases where we should stop speaking about "nature," the possibility of a "we" is coming into existence: What might we be capable of, together, *this* animal and I?[30]

Living Beings and Life

Biologists will undoubtedly find more regarding how pigweed made itself resistant to Roundup. Whatever they find, a thorough reconfiguration of biology is currently underway. Even bacteria are no longer what they were assumed to be—our analogies about them were inspired by a case that today appears particular, their capacity to multiply in the abiotic milieu provided for them in the petri dish or other laboratory apparatuses. Today, biologists realize that approaching bacteria in this way was not paying due attention to the fact that the great majority of bacteria cannot be isolated from the interdependent collectives they require to make their own living. Apparently, the central motif for the history of living beings is no longer the selectivist motif of individual lineages competing for survival. The central motif now would be that of the generation of collectives of interdependent and intertwined living beings, at all scales, each making a living in its own manner, yet with others and thanks to others. If, as Whitehead wrote, life is robbery, the originality that justifies this robbery would have less to do with individual societies and more with modes of composition between societies requiring one another for making a living.

"We have never been individuals," write Scott F. Gilbert, Jan Sapp, and Alfred I. Tauber, with the verve of biologists who understand how much their science has been constricted by the network of analogies that have been established between the individual lineages of biology and the entrepreneurial individuals of the so-called market economy.[31] A new flowering of analogies is now underway that enlarges the imagination of specialists. Known facts are being characterized otherwise. We have known for some time, for instance, that certain termites cultivate fungi that digest cellulose and lignin for them, but when this relation is more fully described and not dominated by the image of the active cultivator taking care of his field, it becomes equally possible to say that fungi cultivate the termites. At every scale, from the cell to the adventure of embryonic development and to the multiplicity of ecosystems, the facts they paid due attention to have led certain contemporary biologists to no longer associate living beings with organisms, or with a distinction between individual and milieu: "Nature may be selecting 'relationships' rather than individuals or genomes. What we usually consider to be an 'individual' may be a multispecies group that is under selection."[32]

The new flowering of analogies does not affect only what Whitehead called nature, however. These analogies can also weld common sense with new imaginations, freeing it from the individualism of "myself, in my opinion" that has poisoned it, and opening it to what it might mean to "make sense in common," *together*, with one another, thanks to others, at the risk of others.

To negotiate this passage among practices implying different types of attention and making different aims matter, I wish to engage with an interspecies example that denormalizes what we have an unhappy tendency to take as a matter of course when we turn to humans: if we explain ourselves well, the other should understand.

When Haraway recounts the manner in which she trained her dog Cayenne in the agility sports they practice together, she knows quite well she will arouse indignation among many of her readers, who will see in it a typical abuse of power by the human over the dog. Indeed this sport involves the dog's capacity to respond to signals indicating to her which pathway to take through a particular course, a course along which she must execute a series of performances that put her to the test, because they are in conflict with any dog's way of making sense of a landscape it runs through. Is not Cayenne's performance demonstrating her blind obedience?

What Haraway recounts, however, is not a story of abuse, even if power is in play. It may be reminiscent of a teacher in mathematics who must transmit a procedure to a student to whom it feels like a perfectly arbitrary way of doing things: Haraway describes how she and Cayenne nearly went crazy, despairing, and losing faith in one other, because the human did not understand the self-discipline this sport would require *of her*. She had to unlearn the connivance and shared meanings involved in interhuman relationships, as well as in her everyday relations with Cayenne. The agility sport turned into traps all the "intuitive" understandings that weave together an everyday life structured through the interchange of habits and expectations. She had to learn not to count on Cayenne's capacity to "grasp" what she was trying to get her to do. She had to prevent herself from generalizations that "for us" are obvious—if Cayenne "knows" this, she should be able to do that. "And so I set about actually teaching what the release word meant instead of fantasizing that Cayenne was a native English speaker."[33]

It proved especially difficult, when a performance was imperfect, to refuse all approbation for the dog, not to offer her encouragement, as one would do with a child, because a child needs it. But Haraway insists: Cayenne is definitely not a furry child. She is adult and different. "I learn such basic things about honesty in this game, things I should have learned as a child (or before tenure in academia) but never did, things about the actual consequences of fudging on fundamentals. . . . Meanwhile, my over-the-top love for Cayenne has required my body to build a bigger heart with more depths and tones for tenderness. Maybe that is what makes me need to be honest; maybe this kind of love makes one need to see what is really happening because the loved one deserves it."[34]

The performance, and what it demands of Cayenne, exposes the hidden economy of communication that prevails in daily life, by imposing a situation in which that economy no longer functions. Cayenne gains the power to assert herself as "other," another who "deserves" nonindulgence, for Haraway needs to address her in a mode that will let her become the competent partner the game requires. Cayenne has to take part in the game, to grasp its aims, and to make it matter for herself.

The creation of "making sense in common" via the game goes through a double transformation: Haraway testifies to both the transformation that makes her capable of loving with honesty and the empirically verifiable change she calls "transfer of authority." When Cayenne, all of a sudden, knows what she has to do and takes charge, it is Haraway who must learn to trust her: no longer guiding her, only giving her signals, which now are anticipated. The transfer of authority means that Cayenne is now in command. She is the one who teaches Haraway the place that is now hers: "Cayenne saw me coming, clipped her smoothly curving stride slightly, and dodged around me, all but shouting, 'Get out of my way!'"[35]

This double transformation is probably what gives its unique meaning to the art of training: the establishment of a relation of partnership whose success is lived as such by the partners who are each made capable of what the other demands. In its contrast with the entanglement of daily relations, training dramatizes this point: before the human comes to understand what the type of partnership needed for the sport of agility requires of her, they have nearly driven each other crazy. Now, however, they share the exultation aroused by the possibility they generated together: a good course, a course in which each in her own mode has felt a connection with the other.

Haraway did not learn by herself "the due attention allowing her to find more" with respect to Cayenne. They learned it together, with each other, through each other, at the risk of each other. What they learned rendered them capable of self-transformation, as partners during a course the composition of which Haraway characterizes as an "ontological choreography," when she reintegrates it into the world of human analogies. In other words, what is dramatized here allows us to leave behind the ensemble of practices calling on nature, and not at all because this relation transcends nature, but because it aims at something other than finding more.

We have the exceedingly bad habit of thinking that "no longer doing science" means indulging in a form of mystic enjoyment. Haraway's account makes clear, however, that the experience of the course is not at all about the so-called oceanic feeling of being at one with the world. To be sure, Haraway evokes an experience of time suspended in a dilated and nondecomposable present, not "preoccupied" by projects born from the past or expectations bearing on the future. Such an experience may well be shared by the top-tier athletes when it is a matter of "giving their all," to evoke a phrase currently in vogue. But it is alien to the attainment of some truth "beyond appearances." It is positively relative to an entirely artificial apparatus, and its success is gauged in overtly conventional criteria. In this sense, it offers a remarkable example of the Whiteheadian difference between aim and goal. The defined goal is part of the definition of the apparatus, but it is not what "animates" the course lived together by Haraway and Cayenne. What animates the course, what the course aims at, is enjoyment of the partnership that the course supposes and provokes.[36]

Characterizing this relation has led me to use terms belonging to Whitehead's metaphysics instead of to his ontology or to societies as they maintain themselves in existence, as they "make a living." Each of the partners belongs to a different species, an Australian shepherd who is the product of years of breeding and a university professor, likewise selected for her excellence, and Haraway stresses that the ways in which each makes a living belong to entangled histories that have transformed the Great Turtle inhabited by Native peoples into the United States of America. And the agility sport also belongs to this colonial history. Nonetheless, the course is also the *occasion* for a kind of experience that, for Haraway, is not explained by these histories, even if these histories situate it: joy. Joy is tasted, it infects its surroundings, it is even addictive, but it is not social in the Whiteheadian sense of character being maintained through a society's adventures.

This is precisely the distinction that Whitehead proposes between "living societies" and "life" in *Process and Reality*. Take, for instance, the thief, who makes his living at the expense of other societies. If the thief were to ask for justification, such a justification cannot simply point out his superior abilities as a thief, even if these abilities are not entirely beside the point. The lineage of Australian shepherds has literally made a living thanks to selective breeding that has intensified its abilities to respond appropriately and enthusiastically to the shepherd's signals. But to successfully make a living does not provide a generic justification for living societies. Making a living is what conditions living societies.

As we have seen, a society, be it living or not, is nothing other than the proposition of a "milieu of belonging" addressed to new actual occasions in the process of becoming. This milieu proposes the actual occasion to determine itself in a mode that prolongs this belonging. These occasions are thus situated socially, but we should recall that this situation does not exert pressure on them to conform. Nevertheless, a society as such, Whitehead writes,[37] is guided in each of its occasions by an inheritance of its own past. It is binding its occasions to a line of ancestry. We may call it a tradition. What justifies a living being's need to make its living at the price of the destruction of other societies, in contrast with a crystal, for instance, is what living societies make possible. Whitehead says that living societies harbor "interstices" or "(socially) empty spaces," which are critical conjunctures where the manner of composing with what is socially proposed may be "original," not conforming to the tradition.[38] In other words, in the case of living societies, tradition is not maintained without opening spaces for originality, for heresy. A living society makes room for experiences "without the shackle of the reiteration of the past." It may give a chance to new manners of inheriting, then binding originality within social bonds, canalizing, or "socializing" it, and becoming itself able to explore new, original manners of enduring. The justification for living societies, Whitehead concludes, is the originality of which they are capable due to the life lurking in their interstices.[39]

Ontology opens into metaphysics here. What Whitehead calls life is not creative power. Originality or nonconformity constrained Whitehead to place the concept of actual occasion at the center of his metaphysics, freeing societies from what characterized organisms: an intrinsic stubbornness whereby self-conservation has its own value. Societies, by definition, matter and make matter, but they may also risk novelty. The justification of the price they make others pay for their living lies in not condemning to insignificance the *causa sui* character of actual occasions, their self-determination of how they will be caused by what causes them. What lurks in the interstices of living societies is what Whiteheadian metaphysics allows Whitehead to think: everything that happens might have turned out otherwise than it has.

Instead of communicating with the dramatization of an experience so familiar that we no longer feel surprised by it, such as the experience of having a body, Whiteheadian metaphysics communicates with what he calls individual concrete fact, which entails the experience of a fact that refuses being referred to reasons more general than itself. Such a fact calls for analogies other than those called upon by the social agents characterized by the sciences. Haraway provides an example of such a fact that makes her experience with Cayenne into a graceful experience in which living being does not (only) make a living, but manifests life—joy is *sheer disclosure*.[40]

We never catch life "in the act" or "red-handed." Life is always what is lurking, and when it becomes manifest, it does so in the mode of a moment that lets itself be recounted while resisting explanation. This is why Whitehead is not vitalist: he does not introduce an opposition between a finalized order proper to living being and an order of blind physical causes. The Whiteheadian concept of life does not propose any model of order. Quite to the contrary, it allows problematizing our usual modes of explanation. We often hear that "it is just a case of adaptation," as if someone had thus explained (away) something that seems new. In response, Whitehead says that a society that adapts is precisely what raises the question of originality.

We must be especially careful not to understand the distinction between "learning from" and "learning with" in terms of the sempiternal opposition between objective knowledge and subjective experience. We must be equally careful to avoid interpreting this proposition as a way to convey by other means Whitehead's proposition: a fusion of nature lifeless and nature alive. Because I wished to resist staging nature lifeless as the lot of science while nature alive would exhibit features directly derived from the metaphysics of actual occasions, I took the path Whitehead had previously opened, linking the term "nature" to that about which we can find more.

Haraway's account of her adventures with Cayenne is nevertheless shot through with borrowings from a biology that never ceases to find more. But this biology is learning in a new way. It has abandoned the search for ready-made, general modes of explanation. Instead, to characterize an entanglement of intrigues, this mode follows and learns to recount it. The problem has changed. The task for biologists is no longer to put in place agents capable of explaining what is observed. The task is to make themselves worthy of what they observe, worthy of what they find.

To characterize the double metamorphosis experienced with Cayenne, Haraway develops an association between the creation of zones of contact, which she owes to colonial studies, and the reciprocal induction of tissues in the embryo, which she owes to Gilbert: "Contact zones are where the action is, and current interactions change interactions to follow. Probabilities alter; topologies morph; development is canalized by the fruits of reciprocal induction. Contact zones change the subject—all the subjects—in surprising ways."[41]

Instead of fusion, a notion like "contact zone" implies communication between disparate registers, all the way from the double transformation whereby Cayenne and Haraway became partners capable of making sense in common, right down to contact between already differentiated embryonic cells generating new types of cells, and passing through encounters between colonizers and indigenous peoples (which makes clear enough that not all encounters bring joy). The contact zone activates a web of analogies whose relevance does not bear on denying what they omit: analogies that enliven attention but do not capture it. Haraway's chapter "Training in Contact Zones" is the harbinger of a nonterritorialized cultural milieu populated with stories, and the harbinger of a sociality weaving exacting, always situated modes of relevance.

The same holds for relations of symbiosis. Instead of implying societies within a milieu, relations of symbiosis place the emphasis on how societies make themselves mutually capable of something they cannot do on their own, thus making their living with one another and through one another. Carla Hustak and Natasha Myers refer to the now-famous case of the alliance between wasp and orchid,[42] which Deleuze and Guattari used as an example of "involution" in contrast with the logics of hereditary filiation that dominate the thinking of evolution.[43] The wasp and the orchid enter into relations of pseudocopulation with each other that defy this logic. This example has received selectionist explanations: the male wasp is said to be "deceived" by the orchid that mimics the genital organs of the female wasp. The orchid is said to trick the wasp in the service of its own reproduction, engaging the wasp in relations that prove sterile for it. Yet, Hustak and Myers object, even if the wasp's relation to the orchid is nonreproductive, why would it not be attractive, enjoyable as such for the wasp? When Darwin experimented with orchids, did he himself not attest to an enjoyment somewhat alien to the unrelenting austerity claimed by the scientist?

While propositions to fuse disparate registers of learning from and learning with would ruffle scientists and render them vulnerable to the trick of evil, scientists may be less prone to deny that their aim, if it is to find more, does not exclude moments of partnership. Such moments are "gratuitous" in the sense that they do not give to what they propose a better capacity to resist objections, yet they subvert the separation between the active role of the one who tries to learn and the passivity required from the one about which learning occurs. At such moments, the articulation between two manners of learning, from and with, may well happen to pose practical and ethical problems for scientists. Biologists would readily attest to this possibility, while for anthropologists such problems are ongoing. Transversal connections—making sense in common—are possible through "life occasions," when distinct lines of inheritance intercompose and intersubvert, from lines of cells and tissues of the embryo and those of symbiotic partners, to the lines of human traditions and the ones that humans entertain with other inhabitants of their worlds. Thus, the solemnity of the world whose experience Whitehead wished to activate might be able to enter into our positive forms of knowledge. Life's justification may subvert the very analogy of the thief as an individual making a living on its own.

{ 5 }
A Metaphoric Universe

In Whitehead's characterization of societies from the atom to the human being, societies are what endure or persist through time, implying a line of continuous inheritance, through a "personal order" in the Latin sense of *persona*. Such a society changes yet maintains a character, which character in turn can characterize the change. Contemporary biology puts another spin on the question, as we have seen. It has discovered that biologists have privileged particular cases, describing living beings in terms of an ability to maintain themselves and to reproduce themselves *alone*, given the necessary resources. The rule, however, is now rather cooperation, or what Donna Haraway calls "sympoiesis."[1] Haraway thus asks, with Thom van Dooren, "who are we bound up with and in what ways?"[2] This question places less emphasis on social persistence and more on partial interdependence among heterogeneous partners, on "ontological choreography," wherein each has need of others but not a need of others in general. To be itself, each needs certain others, and each time in its own partial way.

Welding Imagination and Common Sense

If we recall that the term "character" also designates a fictional character, nothing about this ontological choreography should shock common sense. By the same token, it falls to fiction to give full weight to the question of change, through the problem that a "character" poses for an author. The question of character in fiction does not concern what a character is capable of by itself. It is always about what a character is capable of in a light of how its attachments situate it, generally in the context of a new encounter or under new circumstances, in the mode of "what could it become capable of?" What could it become susceptible to? How would it transform? To keep the reader interested, the author must not subject her character to her sovereign will. She has to generate the experience of "yes, it could become capable of that," which enlarges the reader's imagination.[3]

Along these lines, if we are to avoid a world whose ontological ways are plural only through a degree of originality between the two extremes that are human experience and the conformity in organic societies, we need to relay Whitehead through an activation of imagination that unfurls questions about the circumstances and about what metamorphoses they may occasion. Generality is not an option, neither are *prêt-à-porter* characterizations transcending circumstances. We aim for moments of sheer disclosure, of "becoming manifest," in which it is as if that which we addressed told us that we must now initiate new relationships to understand it otherwise.

This sensation of disclosure does not, of course, afford any guarantee of epistemological validity. It introduces the power of metamorphosis: Cayenne's empowerment is that she does not guarantee. Cayenne has not become able to negotiate, discuss, or argue any more than Haraway has become able to explain what she wants of her. As a dog descended from shepherds, however, she is sensitive to signs, trying to give meaning to them. Thus, when a sign, previously a vector of perplexity, takes on meaning, it becomes entirely her business to answer it, which makes Haraway her witness. And Haraway sharing with us her experience makes us set aside epistemological concerns and territorial conflicts. A welding of common sense and imagination has taken place, so to speak. Specialized forms of knowledge may well, given time, provide different interpretations for the taking-on of meaning that transformed Cayenne. But we will be able to situate these forms of knowledge, to cool their predatory ardor, to call for an enlargement of their imagination. This is because Haraway's story-testimony took the relay of what happened between her and Cayenne, which gives us, we who read it, the ability to make our own connections.

The welding of imagination and common sense is not achieved

in general. It is achieved through intensification, the dramatization of singular experiences, through what might be called "ontological mutations." Sheer disclosure is such a mutation, a manifestation of life, which breaks reiteration of the past, which makes something devoid of importance begin to count, or makes it count otherwise. Whitehead's metaphysics equips philosophy to activate such transformative experiences and to characterize them in a mode such that they resist predation.

Whitehead defined his metaphysics as a matrix for applications. For a mathematician, matrices are operators of a transformation: they thus have to operate; they require something to transform. As such, the application of Whitehead's metaphysical categories should transform what we experience, but experiences do not become mere illustrations of categories. Such categories are tools for philosophy, and philosophers comprehend them only by learning how to handle them, and in which circumstances. This is why to think with Whitehead is to learn. It is to learn to think in zigzag against the straight line. The straight line encourages us to think our statements refer to facts that, correlatively, present themselves as isolable. The zigzag entails experimentation going back and forth between a conceptual abstraction that aims to bring coherence into existence and a situation that our usual statements make bifurcate. The zigzag gives *this* situation the power to reclaim its reality as *individual concrete fact*. "Every science must devise its own instruments. The tool required for philosophy is language. Thus philosophy redesigns language in the same way that, in a physical science, pre-existing appliances are redesigned. It is exactly at this point that the appeal to facts is a difficult operation. . . . Nothing has been defined, because every definite entity requires a systematic universe to supply its requisite status. Thus every proposition proposing a fact must, in its complete analysis, propose the general character of the universe required for that fact. There are no self-sustained facts, floating in nonentity."[4]

Evidently, the aim of philosophy for Whitehead is not to define a fact completely. Neither is it to define the systematic universe that the fact implies in its own limitation. The aim of philosophy is to experiment with language to activate the experience of a fact as individual, whose achievement affirms its own partiality, its "thus." This is why Whitehead ran the risk of drawing an analogy between philosophy and poetry. In both cases, language must call on an experience escaping generalities in order to generate the event of sheer disclosure, which is the only way toward welding common sense and imagination.[5]

The experience of having a body struck me as failing to generate such an event of sheer disclosure, because too argumentative and too impatiently finalized on the conclusion to be reached.[6] The philosopher Michael Schillmeier produces the zigzag, however, when he puts Whiteheadian concepts to work to provide a reading of a situation in which an old lady diagnosed with dementia receives a visitor who takes the time to listen to her.[7] Schillmeier focuses on how the old lady talks about her situation. The message making her experience shareable becomes an ingredient of a painful contrast.

The lady speaks in distress of her powerlessness. She complains of being endlessly confronted with *faits accomplis*; everything is a done deal. She is in a room that has nothing to do with her. Even her clothes are not hers, as if they have been forced upon her. She is, of course, prone to forget, but for Schillmeier, her forgetting does not mean she has lost the capacity to represent her past to herself. What the old lady no longer manages to do is to appropriate the past in the present, to make it *her* past. She can represent it, but it doesn't mean anything to her.

Schillmeier makes use of the resources of the English language—"remember"; "re-member"—to characterize the ongoing activity that such an apparently simple experience requires: *my* room. That experience demands re-making and re-feeling, re-composing, and re-connecting, continually re-assembling heterogeneous elements anew into a continuous composition that *makes sense.* The old lady complains that she is not demented, but she lives a dementing situation, and the term *fait accompli* takes on its full meaning here: the fact is closed on itself; it no longer points to anything whatsoever beyond itself; it is what it is, without any possibility for composition or for "making society." The old lady lives in a sterile and arid temporality, in which what happens to her is outlandish, in which the present, no longer deriving from her past, is subject to a rule of succession that does not concern her; it is just a succession, she says, of done deals, *faits accomplis.* Here, the "it is done" or "it is accomplished" of the accomplished metaphysical occasion finds its caricature, because the "accomplished fact" is bare, without reprising any past or opening into any future.

When the slave Epictetus coldly remarks to his master who torments him by wrenching his leg for amusement, "I told you you'd break it; now it's broken," he doubtless took some pride in this statement affirming the *fait accompli,* for he has made the fact public, and his public statement has effectively become synonymous with letting go, an admirable detachment indicating the slave is freer than the master. Neither letting go nor detachment makes sense independently of its manner. The old lady who lives her very body as de-animated does not let go or reach detachment. She is detached, lives without grasp: panic ensues, as there is no longer any attachment or any point of reference, and meaning slips away.

Yet her tormentors, those who have removed her from her familiar environment, wished only the best for her: she was placed in a retirement home due to fear for her safety. Yes, her dementia certainly worsened, but she would be safer there, where people would take care of her. For professionals, too, there is a *fait accompli*: whatever she does, says, or manifests, she does not give them any cause for doubt; her actions may be placed under the aegis of dementia.

According to the consensual ontology of clinicians, the fact is that the old lady is demented, but her words tell another story, when she is able to speak of her plight with someone who knows how to create with her the possibility of an *entre-tien* that momentarily awakened an experience of "holding and being held" or "making society." Her words bear witness to the manner in which a person may be made demented: de-mented; de-minded.

Yes, of course, in the familiar environment of my office, I can, along with Whitehead, register surprise that I must at once think that I am in this office, observable as a stone block in the middle of a field, and think that the office, in some manner or another, is in me: the office is an element of my present experience, of what I am now. But both the dance of Haraway and Cayenne and the distress of the old lady enrich my imagination in a different mode, making me feel that, rather than being *in* this office, I am, in the moment in which I work, *of* this office. In the first instance, we are dealing with the choreography of two bodies in movement. Cayenne and Haraway have been taught by one another and thanks to each other. Theirs is a body-to-body relationship in which the eyes no longer watch but notice, in which the entire body responds to signs addressed to it. In the other instance, things have ceased to "make signs" for the old lady, which is also to say, "to make memory" and participate in a continuity that would be *hers*. In a conscious manner, the old lady makes evident what epistemology ignores when it takes the changing, variable aspects of our relations with external nature as the primary themes for observation. It ignores the difference between external nature and one's room or den, between space and a *place* charged with memories and histories. Confronted with so many *faits accomplis* through words and things, the old lady is perfectly capable of observing her room, but her room doesn't mean anything to her.

My office is my den, not an external environment in which my body may be found. The books that clutter it are not there only for me to grab; some of them solicit me, silently beckon to me, offer themselves to my attention, and sometimes I get up, somewhat like a zombie, to go look for one of them. They are with me and we share this place.

It is against "dead" abstraction, which would situate me objectively in this office or in nature, for instance, and ascribe all else to an excess of subjectivity, that Whitehead defines what he understands by a proposition: "A proposition is an element in the objective lure proposed for feeling, and when admitted into feeling it constitutes what is felt."[8] Put another way, propositions are that which require what we call abstractions. They are lures that propose a particular manner of making meaning. To admit a proposition into feeling is to experience an articulation: a logical subject (that which is felt) is luring, emerging from a crowd of feelings and claiming this crowd as pertaining to it. When a rabbit bolts, the ensemble of little perceptions making its experience of being on alert is brutally articulated in a vital proposition—"something approaching." When these perceptions are admitted into experience, propositions impact them, and that impact is first and foremost emotional: horror, wonder, disgust, indignation, laughter, or when Cayenne ceases to feel as a *fait accompli* the acceptance or rejection by Haraway of her answer to a sign. She has grasped it, or has been grasped by it, as a proposition telling what the situation requires.

Nonetheless, like other living beings, we do not experience the impact of a proposition if it is socially conformal, if what is felt already belongs, as such, to the milieu that is ours, if it is already socially admitted. At the same time, when Whitehead speaks of "dead abstraction," he does not mean only that it conforms, but also that it appears "normal," shorn of what makes it important. Dead abstraction lays claim to the power of omitting what would situate it. But, as hegemonic in character as this grasp may be, it cannot deny to what it omits the power to make itself felt. As Schillmeier relays the testimony of the old lady in distress for whom everything is now a *fait accompli*, it speaks to us of the catastrophe of a world that is effectively deprived of the effectiveness of "making feel." She can name things, but they are a pure "what." She understands what people say to her, but what they say no longer communicates with a familiar past or a shared future. What she perceives is established in the manner of a matter of fact, without for all that making sense. Unmade is the changing continuity, the intensive strand of variations of interest through which propositions and what they propose to neglect never cease to inter-respond in a mode imparting importance to things felt. It is worth recalling that Whitehead called this strand a living person.

As for Whiteheadian metaphysics, it is a matrix for producing nonconforming propositions that are neither illustrated by familiar, habitual facts nor in conformity with social heritage. If such propositions become objects of feeling, if the statement conveying them is not rejected with indifference in the mode of "that doesn't make sense to me," then they zigzag along with suddenly dishabituated fact, creating an experience of what Whitehead calls *individual concrete fact*. The welding of common sense and imagination, that metamorphic experience Whitehead calls "sheer disclosure," does not involve revelation of some sort of concrete truth beyond our abstractions. Instead it lifts the hegemony that prevents us from feeling our abstractions as living, as engaging in thought, as imparting importance in this way and not some other, and as claiming no power to judge and eliminate what they omit.

The impact of a nonconforming proposition, which makes us think-feel, inspires analogies quite other than those that have prevailed since Hume and that stage a way of thinking by constructing the meaning of mute data, imposing a form on what, in itself, would not commit to anything. The result Humean analogies obtain is pure thought, not engaged in or by the world, which means all the better conveyed by verbal statements that take on the allure of judgments, generating the experience of matters of fact that remain detached from the sense of importance that they nonetheless express. "The notion of pure thought in abstraction from all expression is a figment of the learned world. A thought is a tremendous mode of excitement. Like a stone thrown into a pond, it disturbs the whole surface of our being. But this image is inadequate. For we should conceive the ripples as effective in the creation of the plunge of the stone into the water. The ripples release the thought, and the thought augments and distorts the ripples. In order to understand the essence of thought we must study its relation to the ripples amid which it emerges."[9]

Whitehead here tries to express the type of experience that undoubtedly allowed him to resist doctrines that dismember what we nonetheless know, the experience of struggling to give expression to the impact of a thought (or a nonconforming proposition): the "that's to say . . . ," with no equivalence in play, or "I mean," with no "I" beholding a meaning. With its trajectory of hesitations and reprises as it blindly gropes, thought does not unfold in a rarefied milieu. Thought "makes ripples," which in turn may finally give the verbal articulation allowing at last to say, "and that's what I think," with respect to that which made an impact. What emerges then is not only a thought that has found its expression, but also a metamorphosed thinker who knows what she wanted to say, who inherits the trajectory from which she has emerged.

The image of thought, Whitehead remarks, would have been inadequate if it had suggested a simple relationship. If the ripples were defined only as effects of the stone's plunge, the expression of a thought would seem to find its explanation in that thought. Once corrected, now resistant to decomposition, the image puts it this way: thought seeks its expression, which will render it explicable for and to others, including the thinker herself. It is only then that the thinker becomes able to enter into stabilized verbal relationships with others that make of thought an object on which one may "reflect."

This is, of course, still an abstraction that simplifies the image of the milieu of the plunge. Thought seeking its expression may refer to a solitary episode, but also to a discussion, and in this case, to the suggestions of each and every one involved. Their potential impatience—"come on, express your thoughts clearly!"—might start to interfere, raising more ripples. It often happens that the thread is lost, the process stops short, and a general conversation prevails that leaves somewhat frustrated the person who retains the vague sensation that there had been something of greater importance.

What Whitehead characterizes as a "tremendous mode of excitement" runs in a zigzag with the process of subjective appropriation that is the elementary link of his metaphysics. In any event, the mode of excitement shares its indecomposable character with this process of subjective appropriation, and hence the indecomposable succession of "I mean" and "that is to say" through which a thought seeks its expression. And it also shares the experience of "that's what I think" that marks the relationship of the thinker with respect to what is now "his" thought, with the conclusion of the process of subjective appropriation in the realization of what it aimed for. The expressed thought, having found a satisfactory formulation, has become public, open to being commented on, dissected, evaluated, contested, and reduced, if necessary, to a simple opinion.

The contrast between the private and indecomposable process of feeling one's way and the public mode of existence of a thought that has become an object of reflection, analysis, or requests for clarification dramatizes the contrast instituted within Whitehead's metaphysics between the thick, private, and indecomposable present, which is that of the accomplishment of the occasional subjective process, and what will have been accomplished, now an object available for appropriation by other occasions, data for their own prehensions. What has been obtained has become a heritage for the thinker as well as for others, a public matter, one might say. Still, it is up to each inheritor to determine in a private mode the manner in which they inherit it as well as the value they confer on it.

We may also have the experience of an indecomposable process of participation. One example is the good course that Haraway characterized as ontological choreography. It may equally well happen between humans. A thought may make an impact through repeated, groping statements that interrespond such that a form of shared trance is produced, in a mode that remains opaque to those who listen yet feel it would be an intrusion to intervene, to demand explanations.

Thinking together is rare and precious, marking a contact zone that may be called trust. When Haraway and Cayenne reached the point of making each other crazy, Haraway writes, it was their trust in each other as well as in themselves that was endangered. Here, too, a zigzag may be generated out of the shock that the term "trust" may arouse in some people. Trust is often seen as a renunciation of thinking for oneself, as a letting go that exposes one to the possibility of being duped, being taken for a ride, falling prey to the powers of suggestion. We may know very well that we live, think, and imagine through others, with others, and at the risk of others, and yet, for those who have learned to venerate keeping to oneself, to feel and accept trust is not a matter of inevitable vulnerability, but weakness. It is common to hear confessions such as "I trusted him, and he fooled me" and "we succumbed to his power of suggestion." To have been duped is cause for shame.

If produced, the zigzag unsettles the claims of the "thinking by oneself" to pass judgment on such terms as "trust" and "suggestion." Of course, the zigzag suppresses neither the danger nor the vulnerability (we are at risk from others), but it changes the terms of the question. The ideal of the "by oneself" becomes an incongruity, a shield, as if no thought could plunge into the pond, now frozen.

Trust apparently does not have any direct Whiteheadian metaphysical correlate, any more than do its risks of deception and betrayal. At the same time, the incongruity of the "thinking by oneself" or of the "each for themselves" is correlated to the very possibility of metaphysical concrescence, to the subjective unification of the multitudinous prehensions whereby the subject will determine itself through the way it will make them its own. It is worth recalling how Whitehead placed particular emphasis on not considering prehensions as something initially disjointed, without relation to one another: prehensions in disjunction are an abstraction.[10] Even though they may initially be characterized as a crowd, prehensions cannot be described as a crowd of "as to myselves" indifferent to the aim at unification. Concrescence or composition can occur because, as soon as there is togetherness or grasping together, there is a mutual sensitivity. Composition itself is nothing other than the manner in which this sensitivity takes on consistence, progressively eliminating all possibility of abstracting each component from its relations with all the others.

Nonetheless, this does not mean that each prehension (or feeling) complies with the role that it will have in the composition, as if the role preexisted enrolement. It means that each becomes itself, obtains its concrete reality through the process along which its role is determined. At the final stage, "the feelings are what they are in order that their subject may be what it is."[11] At this stage, mutual sensitivity between prehensions has become relations of fully determined interdependence of each with all the others; each has participated together with all the others in the taking on of consistency by the subject. Composition requires mutual sensitivity, which may zigzag with the experience of what, between living beings, is called trust. This transforms "trust in," whether it is well founded or not, into "trust between," between those who together (engaged by a common aim) strive to make sense in common: not to agree, but to compose through and thanks to their divergences.

In *Modes of Thought,* Whitehead calls the word "composition" a blessed word.[12] And in *Process and Reality,* to characterize the fact that, from the origin of the concrescence, prehensions aim at the subject that will unify them, he speaks of a breath of life animating dry bones, dubbing this initial moment the "miracle of creation."[13] Needless to say, such terms do not bear any particular religious significance. At the same time, they indicate the efficacy Whitehead's metaphysical propositions aim for. While the ontology of societies was dominated by resistance to the bifurcation of nature, and thus was situated by it, Whitehead's metaphysics is constrained by the obligation of coherence and brings into existence what coherence requires. Without composition, there would be no coherent characterization, neither of our experiences nor of our world: disjunction would remain disjunction. And without mutual sensitivity, there would be no composition. But coherence requires challenging all specialized definitions; it demands generic affirmation, which will situate every application of Whitehead's metaphysics.

This is surely why this metaphysics takes on meaning only in the zigzags it induces. Metaphysics can only prove disconcerting to any reading equipped with the tools of the master, as Audre Lorde calls them, that ask us to offer definitions for everything we speak of. Perhaps the children of the master and those of slaves may succeed in creating contact zones, together. In this regard, however, Houria

Bouteldha is surely right to invoke the difficulty of the trust she dubs "revolutionary love," which defies any principle of conservation of the "as for myself."[14]

In Praise of the Middle Voice

For these reasons, we will not define composition, but let it travel where it makes sense, from biology with its zones of contact and reciprocal inductions to what is sometimes called the life of the mind. The life of the mind is not about spiritual life. It is about metamorphic life dealing with the insistence of what since Plato we call ideas that call for realization. As a mathematician, Whitehead is quite familiar with the insistence of mathematical ideas that confer on coherence the power to oblige, giving its life to the mathematical mind. One needs to be alien to the mathematicians' passionate demands to situate mathematics among the master's tools, imposing the authority of their definitions. For Whitehead, mathematical ideas should be radically separated from all authority. In *Science and the Modern World*, he compares the historical role of mathematics to the role of Ophelia in *Hamlet*: Ophelia is "very charming—and a little mad"; she has no grasp on events, and yet she is "quite essential to the play."[15]

It seems to me, then, that there exists a generic characterization of the manner in which nonconforming propositions, a little mad, born of Whiteheadian metaphysics, zigzag and compose with experience: the syntactic twisting that grammarians call the "middle voice."[16] The middle voice stands in contrast to, on the one hand, the active voice, where the subject of syntax designates who, and on the other hand, the passive voice, in which the syntactic subject is what undergoes the action.

Jacques Derrida associated the middle voice with an operation "that cannot be conceived either as passion or as the action of a subject on an object, or on the basis of the categories of agent or patient, neither on the basis of nor moving toward any of these terms"—thus, an operation "that is not an operation."[17] The accumulation of negations suggests the unthinkable that must nevertheless be thought: for Derrida, the middle voice is repressed, and repressing it may be what marks the origin of philosophy.

It is true that our languages, Latin in origin, seem to impose choosing either active or passive voice, either acting or being acted upon. Bruno Latour, however, contributed somewhat to the resuscitation of the semantic pertinence of the middle voice by proposing that we hear it in instances in which we *hesitate* over the attribution of an action.[18] Who acts versus who undergoes remains in question when we speak in terms of letting things happen to us (letting ourselves be led, seduced, interested, taken on board, attracted, recruited, touched, influenced, moved, captured by), and also when we speak of what makes us do something. Such cases are not merely about lamenting or denouncing a flaw, or some sort of lack regarding the autonomy accorded to the subject in the active voice. Sociology itself hesitates, Latour notes. Sociology struggles between the abstract hypothesis of society as the result of individuals acting and the hypothesis of individuals acted on by society. He asks what drops out of these grandiose alternatives. And he answers: the multiplicity of attachments, existences, techniques, and apparatuses that make us do something to others. The question of "who is master?" or "who has the upper hand?" is an empty one. At best, it gives rise to the abyss of infinite regression: "Who pulls the strings of the puppeteer?," and so forth.

Latour proposes instead a careful consideration of the puppeteer's art, the manner in which her gestures respond to the puppet's own movements. The puppeteer's art comes precisely of her capacity to let herself be enacted through *this* puppet, and not to impose movements onto *a* puppet. It is akin to what Haraway calls training in the contact zone, which transforms the subject and all subjects, which makes possible an ontological choreography defying any attribution of responsibility to an author.

Instead of associating the middle voice with a general acknowledgement that we are not the sovereign authors of our actions, Latour proposes to associate it with concern and care over our manners of being attached. Latour's proposal resonates with Whitehead's call: we cannot think without abstraction, for our abstractions are what make us think, but then it falls upon us to remain vigilant with respect to our modes of abstraction. In other words, there is not an insurmountable dilemma here, not a dramatic alternative between an "I think" subject to abstractions that determine it and an "I think" free to gauge its abstractions. From this point of view, vigilance belongs to the middle voice, which implies a cultivated attention toward a possible "change of the subject." We are not vigilant in general, nor can we determine specifically what we need to be vigilant about, for then it would no longer be vigilance, but acts of verification bearing on what is already defined as important. Vigilance implies an indissociable relation between being acted on, which is feeling one's attention attracted *by* something, and acting, which is responding in one mode or another to the question posed by that thing. Typically, each zigzag aroused by a nonconforming proposition takes on the middle voice: the proposition imparts importance to an aspect of experience, but it is up to the activated imagination to "realize" this importance and to explore its consequences, even if it may be in a mode of panic (the trick of evil).

Whitehead became a philosopher to find the means of resisting the bifurcation that dismembers our experience through its demand that either "objective" doings of nature or our "subjective" modes of appreciation are to be responsible for what we know. But then, he gave coherence the power to force him into hand-to-hand combat with the syntax that insists on specifying who acts or causes, who is acted on or caused. This is how Whitehead unpacks the relationship between feelings and the subject who feels them:

> It is better to say that the feelings aim at their subject, than to say that they are aimed at their subject. For the latter mode of expression removes the subject from the scope of the feeling and assigns it to an external agency. Thus the feeling would be wrongly abstracted from its own final cause. This final cause is an inherent element in the feeling, constituting the unity of that feeling. An actual entity feels as it does feel in order to be the actual entity which it is. In this way an actual entity satisfies Spinoza's notion of substance: it is *causa sui*.[19]

Hence, once again, is the importance of speaking neither of goals, which evoke the possibility of an abstract definition of an end pursued, nor of results, which evoke indifferent causes. If we say, "the subject obtains itself from what causes it," however, the middle voice is relevant, for the subject obtains being what it is through causes that have become what they are. But then, in *Modes of Thought*, Whitehead exploits the possibilities of another term. Whitehead specifies that each occasion may said to be "concerned," in the Quaker sense, with things that lie beyond it.[20]

Among all the religious denominations, Quakers are the only one Whitehead cites positively, and in the context of a metaphysical discussion at that.[21] For Quakers, "concern" designates something whose insistence is felt by a member of the community, something that spurs other members to assemble, not to debate or interpret but to devote themselves collectively to *discerning* what is required. The *concern* around which the Quakers assemble does not belong only to a past that the present must determine how to inherit. The present itself is made into a past for a future that will rekindle or reexplore the sense of its limitations. The Quaker procedure of discernment has clearly proved itself, considering that its members have shown themselves capable of political and social discernment for centuries. Yet this procedure does not involve any external criteria of legitimacy. Nor does it have any goals formulated in advance. The question of the future beyond them becomes active in the mutual sensitivity among members of the collective that discernment requires.

"To be concerned by" belongs to the middle voice. To say that they are *concerned* by this realization is one way of saying that prehensions are animated by the aim of which they will be the progressive realization. Mutual sensitivity, to let oneself be affected by the others, is the enactment of a shared concern.

Some experiences are such that we feel the solemnity of "and so it will have been" or "may it not be said that . . ." We feel in such experiences that the finite, the decision obtained or to be obtained, is of importance beyond itself. But the experience can also be of an accomplishment in the present. It is nearly redundant to stress that the indissociable relation between acting and being acted on, between feeling and being felt, between doing and being made to do, characterizes this dilated present, indecomposable, that a good course constitutes for Cayenne and Haraway. The joy with which Haraway writes of how she and Cayenne live a good course together "is tasted."[22] There is not the least guarantee that they experience the same taste. What is important is that Australian shepherd and university professor obtain together and taste together the experience Haraway calls joy.

When all is said and done, taste too may require the middle voice. After all, taste is a matter not only of enjoyment but also of active discernment, and active discernment requires that one agrees to let oneself be affected. It is unfortunate that, like colors, tastes have become symbols of secondary qualities that we are not supposed to discuss. Taste is an integral part of the adventure of life, of encounters that indicate that we are not *in* the world but *of* the world, a world in which it is a matter of discerning between what nourishes us and what poisons us, what heals us (and, if one follows biologists, it is a world in which this "us" includes a bacterial multitude for whom where we are may be a matter of indifference, but not what we eat).

It is worth recalling that touch is a tactile sense, and to touch, one must dare to be touched. Which may make it also worth recalling that Horace's *Sapere Aude,* which Kant made the proud motto of Enlightenment thought, might be translated as "dare to taste" instead of "dare to know." Indeed, Horace wrote, *Sapere Aude, Incipe*—Begin! Kant, of course, omitted the last word. Yet, the final imperative may mean that the exhortation is about the necessity of first taking the risk of letting oneself be actually affected if one wishes to learn to discern. Dare to taste the way in which the situation proposed to you affects you. Dare to taste the mode in which the situation makes you feel and think. Such a motto might prove worthy for critical thinking that does not define itself against beliefs and superstitions.

"Dare to taste" may bring us back to common sense as well, for it is the cooks' art to taste what they prepare. Plato opposed the art of cooks to that of doctors: cooks flatter the senses, he asserted, while doctors, for the higher good of their patients, concoct and prescribe potions of dubious taste. The war machine functioning in the name of reason was launched. At the same time, no one would claim that knowing how to taste signals some immemorial wisdom of the body: it is learned, it is cultivated, it is refined. It even passes through words, as the rich vocabulary of wine tasters attests. But words here do not point, as verbal signs, to the abstraction of categories. Words here stabilize a memory of experiences, activating new contact zones, inducing new sensibilities.

By the same token, "dare" does not signify the flattering heroism of confrontation or emancipation, reason liberating itself from the yoke of seductively reassuring illusions. To dare is first of all to dare to trust, to dare to embark on a journey to learn what is required, to know how to taste, and not to let the injunction of having to prove oneself, or the fear of letting oneself be duped, anesthetize the dynamics that activate the middle voice. *Dare,* but knowing the danger. It is not about pretending to be able to accept everything, but about knowing how to reject, maybe while quaking. Recall that Whiteheadian metaphysics affirms negative prehensions, rejected by feeling: "A feeling bears on itself the scars of its birth; it recollects as a subjective emotion its struggle for existence; it retains the impress of what it might have been, but is not."[23] Reject, but in the mode of "not that, not here, not now," instead of in judgment.

"Dare to taste" does not attribute truth to intuition. Haraway, for instance, had to learn to resist the intuition that suggested to her that Cayenne was starting to understand what she wanted of her. For Quakers as for Horace, "dare to taste" means instead "dare to *begin.*" If you wish to activate processes of discernment through which will be generated a "knowing how" to appreciate it, dare to let yourself be touched. Do not transform yourself into a frozen pond on whose surface the stone will crash, incapable of plunging into the waters and activating ripples across it. Such an image, however, evokes the surely too-particular image of a solitary thinking, inheriting not only from Hume or Kant but also from my own isolated den.

Quakers know how to connect the art of tasting to the collective, appreciating the concern that troubled one of their number, discerning its tenor, grasping its potential importance for the one proposing it, but also recognizing the lack of taste if someone tries to have the last word. For the collective, each word expressing a particular appreciation is always the next to the last. Each such word comes before the obtention of what the encounter of discernment aims for: an agreement that does not belong to any one among the participants. Words then call each other, for the concern has metamorphized, taking on the power to make sense in common.

Put another way, Quakers invented an apparatus of the sort that I previously characterized as generative in the specific context of the *palavra.* It is with such apparatuses that the relevance of Whiteheadian metaphysics, as a metaphysics of the middle voice, is found fully deployed. The efficacy of these apparatuses recalls the middle voice, because it makes the subject what is *generated,* instead of what acts.

Evidently, the efficacy of these apparatuses may be reduced to a general explanation of the psychosocial type, trucking with suggestibility, which will always be in bad taste wherever the abstract ideal of an autonomous subject predominates, endowed with "its" ideas, having to defend "its" positions. One may object that participants know that they have to reach an understanding, and they unknowingly comply with this imperative. Such an objection ignores, however, that American activists adopted the practice of decision-making through consensus from Quaker activists, precisely because it generated decisions that had to hold up when put to the test, especially that of police provocations whose goal would be to divide and sow discord. We must here apply the full force of the word *s'entendre*, to listen to one another and remember that, in most meetings, people do not listen to others; they more or less patiently suffer them. Apparatuses coming from "arts of composition" aim to activate among participants "mutual sensitivity" among diverging voices and perceptions. This sensitivity is not created by the apparatus: the constraints of this apparatus aim instead to struggle against what anesthetizes, against those manners of doing, behaving, and speaking that enclose each person in its aloof "as for myself." Composition doesn't have to be explained, simply cultivated. What is obtained is in the order of metamorphosis: the situation that previously divided has got the power to generate thought and imagination, to arouse the possibility of a making sense in common.

Activating mutual sensitivity is also at stake in the agora as Latour imagines it. The agora, with its assembled specialists, the diplomat, and the public, is an apparatus whose meaning, whose aim, is a possible composition, a composition implying a transformation of the relationship of specialists not to their practice, but to what this practice requires—in this case, a way of linking its continued existence to a certain way of affecting its surroundings. Such a transformation, then, cannot be addressed to humans in general. It can be addressed only to practitioners, to those humans who feel the precarity of what they belong to, who know that the ties of affiliation letting them exist as practitioners may be destroyed or dissolved. It is this knowledge that is intensified in the agora, but in the mode of hesitation. Will I let myself be interested? Will I dare "let go" the manner in which I usually present myself, as serving a cause that should be unanimous?

A METAPHORIC UNIVERSE   { 157 }

Will I let myself be touched by the hesitation of others, by the mode of public attention? And this intensification is possible only if the public knows how to appreciate and taste this hesitation, and knows how to feel the risk of the decision to be made: "amateur public."

Of course, the question posed to each participant in the agora supposes the relevance of the proposition. Latour, as a sociologist, learned a good deal about such relevance from his work concerning human and non-human actors. Relevance here might be said to be a matter of what the experimental sciences call a "crucial" experience, which puts a thesis "on the cross," between success and failure: the experimenter publicly puts a thesis to the test, presenting an apparatus whereby the thesis should be able to defend itself "on its own," without any further transactions, developments, or regulations. There is one important difference: the agora is not a "public laboratory." If practitioners refuse to let themselves be concerned, be troubled by the proposition that is posed to each of them, failure will not be attributable to anyone, for sociologists can explain the refusal through the adherence of specialists to the role that had been institutionally inculcated in them. Here, then, it is not a matter of confirming or denying what the sociologist has "found." It is a matter of responding to a scenario putting each of the specialists "on the cross." Will the practitioner reiterate a presentation of his practice that keeps at arm's length those deemed incompetent, who are supposed not to be able to grasp what her practice makes matter? Will she agree to engage in the test that constrains her to stop being defensive? To be sure, she is free to deny that the scene has anything crucial about it, to refuse to let the situation force her to hesitate, and to keep with her "as for myself." The diplomatic proposition will then crash into the frozen surface of her sensitivity.

The agora's apparatus runs in zigzag with the metaphysics of the middle voice that I have associated with Whitehead's categories. If the practitioner agrees to be concerned by the proposition that concerns her, the ripples raised might reactivate experiences that till then had not been able to make her hesitate, permitting them to get intercomposed with the proposition, to enter into relations of mutual sensitivity with it, to the point where, maybe, the practitioner becomes capable of answering "yes, but" to the one who addressed her. The apparatus of the agora has allowed the proposition of the diplomat to make an occasion. The possibility of exploring together, *with* the diplomat, opens the question "what does the belonging to their practice make practitioners capable of?"

As Michel Foucault taught us, any apparatus may be characterized by its efficacy in inducing particular manners of affecting and being affected. Yet, in contrast to the apparatuses of power he analyzed, the apparatuses I characterize as generative require those whom they gather to be *explicitly concerned* with the question or proposition that assembles them. Generative apparatuses demand that each of those assembled knows that what will emerge from their gathering will not belong to any one of them, but will be the achievement of the "being together" the apparatus brought into existence. The metamorphosis that a generative apparatus aims for is at the same time what is anticipated, what possibility the participants trust in, all the more so if they have already had an experience of it, and what must be obtained again each time. When it comes to a "manifestation of life," no one is in charge of it; it takes place.

As Foucault's work shows through the corrosive efficacy of his analysis, among the many types of apparatus, there are some that need to dissimulate what they aim for behind general justifications, to make it seem as if the type of transformation they induce complied with legitimate needs of society or human nature. Others are rather mobilizing apparatuses that define the milieu as dangerous, endowed with a seductive power liable to incite treason. This may be the case of this institution called Science. Let's remember how, in *Science in the Modern World,* Whitehead emphasized the modern discovery of a method allowing for the training of professionals advancing in their groove and bringing superficial and arrogant judgments to bear on questions insisting outside this groove. For a professional, letting oneself be infected by such questions, which demand they pay attention to what is none of their disciplinary business, would, in effect, be treason. "Don't take a taste of that, or you will be lost for science!"

It is striking how the mode of socialization constituted by this method for the disciplinary training of professionals is analogous to the mode that makes disciplined soldiers. Trained soldiers, however, are meant to "hold together" through real tests instead of mobilizing in relation to an inculcated threat. It would seem that training, punishment, and even the prospect of an execution squad do not offer a sufficient explanation for the transformation of civilians into soldiers capable of obeying orders that may lead them to kill or be killed. In this case the apparatus has for its untold efficacy the metamorphosis of civilians into "comrades," those whom one never leaves behind when confronting danger, unless, of course, the confrontation provokes panic, a frantic unraveling of the band, and then it is a matter of "running for your life" and "every man for himself." The army, then, is not conquered; it is "defeated," undone.

The example of soldiers' comradeship, object of numerous testimonies and matrix of innumerable fictions, is useful for highlighting the necessity of learning to taste what apparatuses do to us and make us capable of doing. If life is manifested through what might be called metamorphosis, few metamorphoses approach the sheen of authority that Cayenne acquires when she grasps what Haraway's signs require. Some metamorphoses come to have a stake in apparatuses whose success is fearsome, channeling mutual sensitivity in an exclusive mode, killing imagination, making what troubles us into a threat. The apparatuses we know the best are those that, everywhere with impunity, prevent, or rather try to prevent, life from manifesting itself, those that aim to eradicate the dynamics of metamorphosis in order to enforce "individuals" endowed with their own reasons to be evaluated according to their own competencies, driven by offers of consumption placing them in the service of growth.[24]

To feel the ontological violence of apparatuses that make individuals of us is to refuse the regime of scarcity judging it normal that some—those who still feel, still think, and still imagine—fall outside the common regime; it is to know that these people are not the chosen or deserving ones, but survivors, partly yet not fully anesthetized. And this knowledge bids us to keep alive the unknown of our era in some way or another: we do not know what humans might be capable of.

Tentacular Affects

This essay has called upon Whitehead to assist with thinking an era of which we do not know whether it marks the end of modernity or explores the possibility of modernity becoming civilized. But it must also attempt to enlarge the field of what spurred Whitehead to think, the field dominated by the triad he associated with the bifurcation of nature (Newton, Hume, Kant). Although such bifurcators are still present and continue to define the horizon of thought, they must become part of the past. And our present, as it is still struggling with them, must become the past to make way for a future that is not defining itself against them. Nonetheless, such an attempt may involve reactivating a past more ancient still. If I began with the stupefaction of Athens' inhabitants when faced with Socrates's nonconforming questions as staged by Plato, it was also because Whitehead once remarked that philosophy was footnotes to Plato's writings. Among these footnotes, there should be one that would raise questions about the apparatuses that ensure that, from our school days, we know that we must answer Socrates's sort of questions if we wish to be heard.

Whitehead wrote, "the account of the sixth day should be written, 'He gave them speech, and they became souls.'"[25] This does not mean that speech endowed us with souls. Speech may make us more subject to conformity than a cat or even a rabbit could ever be, capable of enforcing and justifying highly noxious modes of abstraction. The soul in Whitehead's sense is an outcome that he links to the individual concrete experience he names "disclosure." The experience of disclosure, in which life is made manifest, has nothing exceptional in itself, but when we lose the *sense* of this experience, when we attribute elucidation to ourselves as a result of our own activity, when we forget to give thanks to what has made it possible, "we are shedding that mode of functioning which is the soul."[26]

To become a soul, then, is also to become susceptible to losing it. Speech allowed Socrates's nonconforming propositions to find efficacious verbal expression, but it also allowed them to be made into a requirement demanding our conformity to it, which may pose a threat to souls. In *Adventures of Ideas,* Whitehead associates the ambivalence of this particular efficacy with the entities Plato called "Ideas."

> But the notion of mere knowledge, that is to say, of mere understanding, is quite alien to Plato's thought. The age of professors had not yet arrived. In his view, the entertainment of ideas is in-

trinsically associated with an inward ferment, an activity of subjective feeling, which is at once immediate enjoyment, and also an appetition which melts into action. This is Plato's Eros, which he sublimates into the notion of the soul in the enjoyment of its creative function, arising from its entertainment of ideas. The word Eros means "Love," and in *The Symposium* Plato gradually elicits his final conception of the urge towards ideal perfection.[27]

And he immediately adds, without commentary: "It is obvious that he should have written a companion dialogue which might have been named *The Furies,* dwelling on the horrors lurking within imperfect realization."[28] It is important to recall that Ideas, here, are not another name for propositions. Ideas are born in the particular milieu constituted by Athens, where young people competed in argumentative jousts that earned them the name "Sophists."[29] Plato enrolled Ideas to differentiate the philosopher from all others: the philosopher has access to Ideas, while others play with words and profit from false resemblances. Assigning to philosophy the task of vigilance toward our modes of abstraction, Whitehead himself pens a footnote to Plato's text with reference to the missing dialogue warning us about the formidable power of Ideas: Ideas may also make Furies of us. Still, it is not surprising that Plato never pursued a dialogue on this topic. As baptized by Plato, Ideas had already lent themselves to realization as so many weapons of war against those who stood accused of utilizing speech for seducing us or fooling us. Poets included. Clearly, Plato would have identified Whitehead, who spoke of philosophy being like poetry, as a Sophist.

Still, for better and worse, ideas have continued their adventures, and like Whitehead, we have learned that the idea that activates may equally well devour, that to be touched by an idea is also to risk being possessed, becoming prey to it and losing the mode of functioning called "soul." Consider professionals, for instance, who are incapable of thinking before the people who will pay the price for the abstraction that is feeding on them. To be realized without becoming furious, the idea necessitates being *entertained* instead of enrolled. Whitehead often uses the term "entertainment," be it for ideas or for possibilities that ask to be realized, and it is worth recalling that, etymologically, this term refers to the art of hospitality as well.

As nonconforming propositions, Ideas are neither good nor bad. They have the power to touch us and to solicit a realization that goes by the name of thought. The dialogue that Whitehead laments for not having been written would have dealt with the necessity of tasting, of discerning, as the Quakers knew how to do, remaining vigilant with respect to the manner in which we let ourselves be touched and the manner in which such touching makes us think. As regards the touching and being touched associated with the middle voice, Haraway inspired me to think in a tentacular manner.[30] Its tentacles make the octopus especially sensitive to its world as they palp, seek, and explore, and yet tentacles may capture and suffocate as well.

The idea touching the mathematician embarks him on an adventure. But the idea whose realization is associated with the power of combating confusion and false resemblances spurs an anesthetizing or even murderous mobilization. It is doubtless why the metaphysical propositions of Whitehead function in zigzag, activating thought but never giving it the power of defining, of claiming possession of what is proposed.

The question, then, is no longer only about what Whitehead called civilization, nor about that universal called the human individual, who claims to have a soul. As much as anthropology, biology today asks us to consider such an individual as a particularity. For biologists, what we judge to be "normal," an individual organism in its milieu, is anything but obvious. And anthropologists, when they step out of territories conquered by modernity, rather deal with *persons*, who are to be characterized in the manner of what I have called an "obtained," whose proper value is inseparable from the links and alliances that situate the person and the obligations involved by these links and alliances. A tentacular version of Whitehead's personal order might prove relevant here. How one becomes a person is a question that concerns us as much as it concerns peoples who have cultivated it, but in different modes. We have defined ourselves against these cultures. We have accumulated practical aporias of which we are proud, all of which hinge on the enigma of "making oneself by oneself." What if taking seriously the tentacular character of what "being a person" requires involved a culture of entertaining beings (as I will call them) without characterizing them otherwise? A culture where the manner in which we can characterize them is relative to the relation, and to the metamorphosis this relation requires.

The words of a Native American of the Omaha nation entertaining a boulder, which, for the moderns including Whitehead, is incapable of hearing them:

> unmoved
> from time without
> end
> you rest
> there in the midst of the paths
> in the midst of the winds
> you rest
> covered with the droppings of birds
> grass growing from your feet
> your head decked with the down of birds
> you rest
> in the midst of the winds
> you wait
> Aged one[31]

The one who speaks thus addresses a boulder as an animated being, in a mode we should not hesitate to call animist. Now, animism, in our civilization, is synonymous with adherence to frankly obsolete beliefs. If the task of philosophy, in Whitehead's sense, is not to transcend the civilization it is part of, can philosophy nonetheless problematize the judgments of our civilization? There, where humor's disclosure of the absurdity of bifurcating doctrines has no efficacy, might not the middle voice I have associated with Whiteheadian metaphysics crack the walls that exclude what, for us, obviously has no soul? Can we let ourselves be touched, be engaged to feel and think, by the words of this Native American?

When art is in question, a stone statue for instance, we know that "letting ourselves be touched" is what is asked of us. When confronted with pain, disappointment, or discord, we know that the manner in which we make our response to what, irrepressibly, touches us, engages us in a risky history, as is attested by professionals in care, for instance, who deem it necessary to conserve a stone-deaf emotional distance, to remain in their role. So be it. But it is in this way that they may become tormentors, justifying their distance and their role through theoretical judgments: "In any case, she is demented," or maybe worse, "she is setting a kind of trap for us, unconsciously." Here, it is about letting oneself be touched by a mode of relation from which we are usually "shielded," with no need for justification. The judgment "this Native American is animist!" can at best be modulated by a certain nostalgia, a vague poetic empathy, yet will preclude any doubt: the boulder is indifferent to the words of the Indian.

To be sure, it is not *a* boulder, a "block of stone," the Omaha addresses, but *this* boulder, in *this* place where it rests, of which it is an integral part, immemorial, inseparable from other beings with which it shares the place: winds, birds, grasses, those who pass along the trail. To imagine a bulldozer, for which it would be nothing more than a block, is to imagine the destruction of this place, a *fait accompli*, to adopt the turn of phrase of the old lady deemed demented. Such a fact counts those beings attached to the boulder as nothing. While this first uneasiness may well arouse a sense of remorse and scruples, it may also provide reassurance a little too cheaply. Contrary to the old lady whose removal from her home has rendered her "demented," we ourselves are capable of changing place, of reweaving attachments. We know that it is "merely" a matter of habits. The notion of traditional cultures takes a similar tack: the respect we pay to them now and then signals their fragility and their dependence on conservation values. This also means that the choice of safeguarding belongs to us, tolerant humans, who deem ourselves at home wherever we are.

"It matters which ideas we think other ideas with."[32] Against the Furies, let's adopt Haraway's phrase as a talisman that forces us to think with the consequences of our ideas. The question, then, is not only to let oneself be touched but also to make oneself capable of answering for the manner in which we think the relation of thought that the Omaha Native entertains with the block of stone. It is not by chance that Haraway owes this phrase to Marilyn Strathern, an anthropologist. Other contemporary anthropologists, such as Eduardo Viveiros de Castro, Helen Verran, Marisol de la Cadena, and Lesley Green, have given new significance to the task Whitehead assigns to philosophy: not to explain by eliminating, by depriving experience of its proper value. They have made the opposition—they who believe versus we who know—into a professional flaw. They have also refused any passport permitting them to feel at home everywhere, or to assert that "nothing that is human is foreign to us." What Viveiros de Castro characterizes as the task of "decolonizing thought"[33] involves then resisting the temptation of what Verran calls "anthropological bad faith."[34] Here it would mean deliberately depriving oneself of all resources that permit interpreting what the Omaha Indian said in a mode that appears respectful but in fact comes down to *denying authority* to this Omaha Native over the nature of his engagements with the world.

Neither is the point to bow down to that authority. We are not dealing with experimental science, with reliable testimony about the manner in which a boulder ought to be defined. At stake is a rigorous practice of critical imagination: if the thinking that thinks the thinking of the Omaha Native lays claim, in one way or another, to the power to situate the Native American, to "understand" him better than he understands himself, it must be carefully, painstakingly, conscientiously problematized, not to pronounce it "guilty," but rather "trivial" in the mathematician's sense in order to indicate a failure in formulating a problematic: one finds what was already known.

To let oneself be touched is not in the least to try to transform us into a Native American, but to try to confer on him the power to situate us. We are dealing then with what William James called a voluntary act, and it is in this way that James distinguishes between imagination and imaginary. The fundamental act of Jamesian will is not the decision taken by the subject; it is the tension of attention, the effort to give importance to a thought that is a priori unfortunate, intrusive, or running counter to our habits, to hold it "fast, in spite of the host of exciting mental images which rise in revolt against it and would expel it from the mind. Sustained in this way by a resolute effort of attention, the difficult object erelong begins to call up its own congeners and associates and ends by changing the disposition of the man's consciousness altogether."[35]

James's description could be an application of Whiteheadian metaphysics in its tentacular version, calling for a thought in the middle voice. The effort to "not expel it" implies an aim to obtain instead of a goal to attain, a response to a subjective insistence instead of a decision to accept. In this effort, we can detect a concern for what lies beyond ourselves, which Whitehead associated with Quakers. The present, in which we prove incapable of hearing the Native American, becomes the past for a future that will reenact the sense of this limitation and reenact the manner in which the judgment "it is merely animism!" will be received: other thoughts will think this judgment; it will not be forgotten, but problematized, separated from the furious power of differentiating thought and belief.

This does not mean the sad relativism of "to each his own mode of thinking," a relativism without effort, ironic, an avatar of the bifurcation of nature. We aim at acknowledging in the Omaha Native the power of situating us. This power is not about making us recognize that he knows better than us what a boulder "really" is, for it is we who cannot prevent ourselves from doubling a real feeling with the question of what it is the feeling of "really." We do not taste experience, but pass from the experience to the question of what this experience is or is not the reliable witness of. The Native American does not ask such a question. Instead, he will doubtless wonder what comes over us when we pose such questions. And therein may indeed lie the tentacular version of the question, the one that touches us and forces us to think: What comes over us? What has come over us? What thinking makes us think the thinking of others in a mode that makes of us masters who attribute meaning to our experience and to theirs? What gives our ideas their furious power?

What has come over us? The manner of formulation matters, because we must remain suspicious of modernist fables in which a war is waged between the celebration of the greatness of Man and the denunciation of his Guilt. These *prêt-à-porter* thoughts never seem to fail to give us trivial answers that confirm our eminent responsibility and give others only the role of victim, even while dressing these victims in the innocence we have lost.[36]

David Abram makes a proposition of great interest by approaching the question of what has come over us in the mode of "intrigue" instead of "sin." His point of departure touches on an exceedingly sensitive point for me. In his struggle against Hume's empiricism, Whitehead found in the body the most intimate experience of what his metaphysics calls for. But he abandoned perception to Hume, associating it with the triumph of abstraction by placing "external nature" such as it is perceived under the sign of a radical asymmetry between "I perceive" and "it is perceived." Counter to Hume, who denied that it is "with eyes that we see," Whitehead evoked oculists and antiprohibition leagues. For his part, Abram practices the ancient art of prestidigitation, and this is what allowed him to forge connections with indigenous shamans who were appreciative connoisseurs of his practice.

"What has come over us" is what spurs us almost automatically to ask, "is it true or is it an illusion?" If shamans and the prestidigitator share a bond, it might well be forged in the refusal to transform this question into tragedy. The practice of prestidigitators is certainly an art of manipulating the senses, of taking advantage of perceptual abstractions and the anticipations they activate, and yet, if they can take this advantage, it is because the senses are engaged *with* things, as certain neurocognitive scientists now recognize.[37] To seduce and to be seduced, to attract and to be attracted, to solicit and to respond to a solicitation—according to Abram, such reciprocity demonstrates that the sensory apparatus is "tuned for relationship" instead of disengaged observation first and foremost. We must take seriously what we say: "This boulder attracted my attention." The sensible object is not available for grasping; it proposes, promises, invites, solicits, induces, and may even capture.

> Each presence presents some facet that catches my eye while the rest of it lies hidden behind the horizon of my current position, each one inviting me to focus my senses upon it, to let the other objects fall into the background as I enter into its particular depth. When my body thus responds to the mute solicitation of another being, that being responds in turn, disclosing to my senses some new aspect or dimension that in turn invites further exploration. By this process my sensing body gradually attunes itself to the style of this other presence—to the way of this stone, or tree, or table—as the other seems to adjust itself to my own style and sensitivity. In this manner the simplest thing may become a world for me, as, conversely, the thing or being comes to take its place more deeply in my world.[38]

In the texts of philosopher Maurice Merleau-Ponty, Abram finds inspiring testimonies for a reciprocity between feeling and felt, seeing and seen, touching and touched, a reciprocity I would call "tentacular." And, far from urban life, he himself had the transformative experience of relationships of coanimation defying the dismemberment between active and passive voices: like an anesthesia that would be lifted, a relationship that would be regenerated, senses that would be awakened, and above all not effects of belief "being projected" on a world that, for its part, would remain what it is, passively offered to our point of view.

But then, if our sensory apparatus, being participatory, makes "animists" of us, what has happened to us to make us so proud of not being animists? How are we to recount the history that has allowed the statement "perception is the triumph of abstraction" to feel plausible? In this respect, instead of recounting the grand history of a progressive disenchantment that separates us irremediably from our origins, the great interest of Abram's hypothesis lies in proposing that we experience a surprising realization. Animists we were, and *animists we always are,* and now more than ever. Participation or attunement *with* things, whereby they become animated and animate us in return, has never been interrupted, and cannot be, but its site has changed. What "came over" us would be a new relationship of intense coanimation that surged up between the sensory apparatus and alphabetic writing, the only writing that allows words to be imposed on us in an irrepressible manner, as self-sufficient, as meaning something.

> In learning to read we must break the spontaneous participation of our eyes and our ears in the surrounding terrain (where they had ceaselessly converged in the synesthetic encounter with animals, plants, and streams) in order to recouple those senses upon the flat surface of the page. As a Zuñi elder focuses her eyes upon a cactus and hears the cactus begin to speak, so we focus our eyes upon these printed marks and immediately hear voices.... *This is a form of animism that we take for granted, but it is animism nonetheless—as mysterious as a talking stone.* And indeed, it is only when a culture shifts its participation to these printed letters that the stones fall silent. Only as our senses

transfer their animating magic to the written word do the trees become mute, the other animals dumb.[39]

Thus, according to Abram, we the literate would be animists. And so would be the other humans who inhabit our so-called civilized world, in which the signage is everywhere, indicating what to do and where to go, in which lights turning red call for us to brake and ideograms on doors advise us not to go into the wrong toilet. We live in a world that lets itself be read, but we cannot feel surprised by it, because texts and signage are perceived as expressing an intention, that of their author or of someone who put the signals in place with the intent of something to tell us. We live in an urbanized (civilized) world, which is also to say, talkative, saturated with intentionality making us do things, making us feel and think. It is precisely this intentionality, this meaning, that we cannot attribute to a boulder. With good reason, for that matter, because, even for us, the question "what is your intention here?" is not neutral, and sometimes becomes as aggressive, as "what do you mean by this word you are using?" The question of a confessor or an inquisitor. As Alice might say, how can I know what I mean to say before saying it? And Whitehead would agree: aiming at is not the expression of an intention.

Whitehead noted that "the effect of writing on the psychology of language is a neglected chapter in the history of civilization."[40] He then expands on this point: "Of course, we are much more civilized than our ancestors who could merely think of green in reference to some particular spring morning. There can be no doubt about our increased powers of thought, of analysis, of recollection, and of conjecture. We cannot congratulate ourselves too warmly on the fact that we are born among people who can talk about green in abstraction from springtime. But at this point we must remember the warning—Nothing too much."[41]

Nonetheless, the very example Whitehead selects is still "lettered." The contrast he evokes is ours. To let ourselves be situated by the words of the Omaha Native, we must cease speaking of the enhancement of our capacities by writing, singled out from among all the devices through sympoiesis with which we have become what we are. We must stop forgetting that, on a lovely spring morning, "non-civilized" humans doubtless see something quite other than "green."

It is not a question of proclaiming writing to be nefarious, even guilty, but of learning not to mistake its power. As a philosopher, I am a daughter of writing: one becomes a philosopher through encounters with the writings of other philosophers. Yet it is not the theses advanced by other philosophers that secure such a becoming. Instead, when there genuinely is an encounter with certain philosophers, it is through the setting into motion of feeling and thinking by what has forced them to think, by what has made philosophers of them. It is much the same for mathematics, for literature, and for so many other practices born of the written that have transformative potential. They may, as Gilles Deleuze wrote, "make larva of us,"[42] which is to say that they may unmake the habits and intentions of an "I" that is then no longer "in the world," or in her den, but is captured, transfixed, "initiated." In brief, they expel the one who reads from her position as a reader in the active voice, who deciphers what the author wishes to make known to her; they instead affect her in the mode that I have associated with the middle voice.

We are no more guilty than writing in itself is harmful. It may be, however, that the Omaha Native would call us ungrateful. No other world has ever made its human inhabitants as dependent on a multiplicity of others as the modern world has. It may be that no other world has pushed ingratitude, and thus imprudence, toward others so far, including the other that is writing; no other has cultivated deafness toward the manner in which others solicit us to such a degree; no other has transformed what we owe to relationships with others into property rights (my past, my sensitivity, my thought, my interpretation of this text). In *Science and the Modern World,* Whitehead underscored the importance of an education that cultivates habits of aesthetic appreciation. We should also think of an education that cultivates "entering into relationship" and metamorphoses that name and honor those with whom we have made relationship, and those who know how to let us taste how this relationship engages us.

So often we ask, "do we have the right?," only in order to grant to ourselves the authority and permission to ignore the consequences. If we cultivate a sense of being obligated, we undo the type of security we ask of "the right." An ontology of tentacular lures communicates with a sense of existential precarity that calls for what James described as "consent," accepting the risks inherent in the tentacular capture instead of counting on a world in which things patiently conform to their role and we to ours.[43]

Living in the Ruins

Haraway proposed an important analogy between the string-figures game and her practice of thinking and working.[44] String-figures games are found nearly everywhere across the world. Several players are needed for this game; one does not play it alone; one does not come up with ideas oneself. At each stage of the game, a string figure is held out by one player, which serves as motif for the next player who takes it up. The first player lets go the figure she held as the next player takes over it, producing a new figure. Figures are successively made and unmade. The player who holds out a figure remains passive while the next responds to the offer, engaging with the interlaced strings to deploy a new figure. Each new figure is at once deployed and exposed, open to serving as motif for another player. The play may recall the indecomposable chain of the actual occasion of Whitehead. It also proposes a figuration of "life": instead of staging the persistence of a composed figure, the play stages a tentacular becoming through the passage of relays making each string figure into a motif or proposition, generating continuity that is also metamorphosis.

My attempt to think a tentacular version of Whitehead's metaphysics does not contradict the ontological figures Whitehead lays out. It takes up the strings held out by his metaphysics to produce different ones. Whitehead situated every nascent occasion in relation to a proposition of social conformity. Of course, his metaphysics denied neither social entanglement nor interdependency among societies. In fact, in *Science and the Modern World*, he took issue with the successors of Darwin for having ignored how the cooperation among organisms mutually produces a favorable environment.[45] Nonetheless, in *Process and Reality*, he continued to speak of philosophy of the organism, which provides a good indication of the importance he gave to the success constituted by the "holding together" of a society. From the atom to animal societies and human societies, those who compose the society reiterate its way of holding together thus and not otherwise. As for the milieu, it is mostly qualified in terms of patience with respect to this reiteration. In contrast, when Haraway speaks of "worlding," what matters are the *with*, the *through*, the *thanks to*, and the *at risk of*: a world is never *my* world; it is that entangled world in which I am actively engaged, touching and being touched, a world that solicits me, proposes to me, "motivates" me. It is that world at whose risk my manner of living and dying belongs.

There is no contradiction between organismic and tentacular versions. There are different ways of making things matter. Hume, Newton, Kant, the biology of the organism, and the decline of civilization generated a motif for Whitehead's thought. In my attempt to reprise what he proposed, I have seized upon it with another aim. From that past, I inherited not only the defeat of common sense but also the worldwide destruction of generative apparatuses. If I have conveyed what Whitehead proposed for thinking less as some kind of irrevocable defeat than ongoing devastation, it was not by sovereign choice. In fact, it is in this way that I let myself be touched by him. If devastation is ongoing, so must be resurgence. In fact, this is what I experienced reading: we may *take* the right to think that! It is an experience of *regeneration* that I wished to relay, and which I ultimately characterized through the figure of the middle voice.

This leads me to put quotation marks around the term "modern civilization." Modern civilization has worked for the destruction of apparatuses cultivated by peoples for whom language was not reduced to a tool for intentionally communicating what we mean to say. Instead, language participated in the fabrication of always precarious worlds, ceaselessly reprised. Indeed, I reprised Whitehead in a manner that is not "mine." "My" reprise was activated by the situation that is ours today. It might be said that the question of the decline of our civilization has given way to the question of its collapse. Cracking and grinding can be heard that indicate the calving of ice sheets. Grounds once considered secure are dislocating.

Cultivated academics announce that we have entered the Anthropocene, the Age of Humans who discover that They have impacted the world in such a mode that the history of the planet itself is affected. Amid the destruction of so much, worthy of note is the activation of indomitable entangled dynamics that are likely to destroy the habitats of innumerable species. It would seem that the Age of Humans is, in fact, about to come to an end.[46] Nothing but tired myths come in response to the devastation. For instance, humans discovering their de facto responsibility will rise to take responsibility and succeed in mastering the uncontrollable (geoengineering). Or else, taking leave of their ravaged birthplace, humans will migrate to other planets to be terraformed. Or they will succeed in decoupling their civilization from the material and energetic flows on which it depends and will live off the ground thanks to technological miracles. Or else again, it will be the end of the world, punishment for our hubris.

It is nonetheless true that, in only a few years, climatological models have allowed "finding a lot more" about what has been called nature. These models have multiplied the number of agents whose potential sensitivity to each other implies consequences that have forced researchers to quickly rethink all the previous abstractions that implied the perennial stability of the Earth. But finding more, even explaining what is happening, has little to do with becoming capable of responding.

On the contrary, climate models can stupefy us with their abstraction. Thus, one often hears that, according to their calculations, in 2050, it will be too late to attempt to avoid the worst. It is as if the moment of truth were at hand, the verdict condemning us. We forget that other peoples have already, at our hands, lived through untold devastation, without any truth other than that of the destruction of their world.[47] And we also forget that, even if we proved capable of escaping the worst, our descendants will go on living on a poisoned and exhausted earth in a profoundly and *very* persistently disrupted climate, an earth from which a great part of living beings will have disappeared. Thus "2050" does not mark the closing of the curtain, the end of the play. What is coming will indeed demand learning to "live in the ruins," to quote Anna Tsing.[48]

For many people who greet the idea of the end of the world with equanimity, the thought of going on living in the ruins is intrusive in James's sense. A deliberate effort is needed for the mind to hold onto a thought to permit it to evoke "congeners and associates" who can give it consistency. My attempt at a tentacular version of Whitehead's metaphysics contributes to this effort. It does not give a crucial importance to the possibility that, in 2050, or even today, we must conclude that it is too late. We have to think and imagine before those who are already living in the ruins or will tomorrow live in the ruins, before the innumerable living beings, humans and nonhuman, who, in one manner or another, will continue to live and die on this earth, regardless of our conclusions. And this creates what James called a "genuine option," an option not possible to avoid on the pretext that the game is over. In one manner or another, the inhabitants of ruins will inherit what we leave them. As in the string game, we are engaged by the figures that we will prove capable of proposing.

We are indeed no longer in the streets of Athens. Nor are we in the violent era in which Leibniz dreamed of a philosophical reason capable of curbing the furious arguments that justified war and massacres, with his *dic cur hic*, or "say why here." Instead of the discourse of those who know, it is the situation itself that defies common sense. It would feel vaguely obscene to ask a robber who steals the chance for a future from innumerable living beings to find any sort of justification. To opt against the set of perspectives and histories that justify our civilization is not to show that it is all wrong. To opt for learning, right now today, to live in the ruins is to opt for learning to think without the security of our proofs and to consent to a world that has become intrinsically problematic.

Are we not quite far from common sense that broods? According to Abram's hypothesis, we may not be all that far. For Abram, the torpedo effect of the Socratic method lies in the stupefying character of questions that are off the ground, beyond circumstances, detached from memory. School children, those who will live in the ruins, may be similarly torpedoed: yes, to be sure, to pass from the practice of dividing a pie into three slices to understanding what the fraction ⅓ means can awaken us to a new world. Yet such understanding is not about the manifestation of a capacity for abstract thought in opposition to concrete knowing-how. It is a matter of the potential success of a genuine metamorphosis. Such an operation is always full of risks, however, because the child who does not understand anything and the teacher who does not understand that the stupefied child cannot understand can, like Haraway and Cayenne, drive one another crazy. The Greeks held mathematics to be the very example of the transmissible, something that, once disclosed, is self-evident. But this is because, as with writing, entering into relation with a mathematical being entails the risk of an amnesiac capture, what today we call "competence."[49] But what is acquired? We say, "I got it." But we might equally well say, "it got me."

Still, learning to live in ruins is not about common sense in the sense in which Whitehead understood it, the common sense of civilized humans who were counting on things (or on slaves in Athens) functioning for them, without them. Other than the absence of guarantee and the right to count on, nothing else defines what "ruins" means here. Whoever speaks of ruins now speaks of an apprenticeship in the art of attention in a world that no longer conforms to the roles assigned to it by our habits. In this world, nothing is self-evident and nothing happens by right. Consent to precarity must be cultivated, but obviously such precarity has nothing to do with the obscene precarity produced by the politicoeconomic imperative of generalized flexibility.

The art of attention is an art of the middle voice, a tentacular art, for it is about allowing oneself to be touched and conferring on what touches us the power of making us feel and think, always *here* and never *off the ground*. Might we not then recognize the artful rites and initiations belonging to "traditions" that Whitehead assimilated to the maintenance of social conformity as creating a contact zone with the ground that generates an art of attention, as what we must care for if we are to receive from this ground the capacity to cultivate ways of living and dying that are not reducible to mere survival in the ruins? Care for entangled strings, care for motifs that strings compose and recompose, care for the manner in which these motifs hold the situation together—learning to live in the ruins is learning to "make sense in common" within a tentacular milieu wherein no signification or convention can be counted on or taken for granted.

"Making sense in common": a sense that the common situation is problematic not only for humans but also for the ensemble of what participates in this life in the ruins. Naturally, the objection may be raised that problematizing is a human art. Undoubtedly, that is the case if one understands by "problematizing" an examination of the legitimacy of a position, putting one's own reasons to the test and calling others into question, in short, behaving as a citizen speaking in the agora or as a philosopher writing a critique of a colleague's arguments. But the arts of composition teach us something else: they teach us that the discursive regime designed to determine a winner and a loser is precisely what must be avoided. From the practice of *palavra* to the procedures invented by American activists to create what they call consensus among them, these arts teach a feeling-together of the manner in which a situation affects each person. The arts of composition are arts of slowness, as once a statement is articulated, they exclude all interpretations that fall back on the intentions of the enunciator and what they "mean to say." No one is supposed to defend themselves or to contest or strive to uphold a signification. It is not about courtesy here, or if it is, then it is courtesy toward the statement itself, which must be welcomed as belonging to no one, as being issued from the situation, and its reprise by others must be seen as contributing to the composition of the situation.

The art of *palavra*, one could say, is a collective and situated art of problematization, activating what Whitehead would call "the compulsion of composition."[50] We can speak of an operation of composition without a composer, and of course, without a position of transcendence allowing us to evaluate what has taken on reality. The only criterion is immanent. It refers to "knowing how to taste" of those who are transformed by their participation, transformed but not converted, for the composition is not an agreement that transcends divergences. It is about the situation that obtains the attention it calls for. Composition is about a transformation of divergences as so many dimensions entangled by the situation itself, to which it is a matter of consenting without dreaming of reducing it to a problem that imposes the terms for its solution.

From the ontological point of view, then, what makes sense in common would no longer be the maintenance of social conformity, but the continuation of the composing Haraway calls sympoiesis. In fact, with the notion of proposition, Whitehead already opened this pathway. We need only recall that Whiteheadian propositions are in a contingent relationship with human language and propositional efficacy is above all in the order of a lure to feeling. The proposition is proposition for sense-making/feeling-making/motif-making. If it is admitted into feeling, however, it does not say how it will be felt, or what signification to give to what it makes feel, or what it will motivate. Propositions are vectors of partiality. Although any relationship that mobilizes ways of making something count or making something matter requires propositions, propositions do not determine the signification of the relationship. Belonging to a common, precarious composition, always without guarantee, entangling relations always to be determined, they do not involve a convergence that would enroll and unify. Interdependence, the *with* of sympoiesis, is not a cause to defend. As is the case with social endurance, interdependence is in the order of "fact." The fact is that each partner functions as a proposition or motif for others, from one to another or in cascade, *but always each in their manner.* In contrast, what can be defended against every appropriation is the question that is the unknown of every composition: "What is belonging capable of?" It is the question that aims to cultivate the arts of composition.

We often beg "to be heard" and not "to be comprehended." In other words, we are not asking "listen to me," but "let yourself be changed." Such petitions typically implicate the middle voice, for they address their appeal against the resistance to hearing, the refusal to taste. The collective practices activated by generative apparatuses are addressed to the same unwillingness to hear. In either case, success puts into play the capacity to let oneself be affected or inflected, not to be persuaded. In *Adventures of Ideas,* Whitehead celebrated the power of ideas, which implied persuasion and not force. Can we speak here of "ideas"? Certainly not, or at least we cannot speak of ideas insofar as the constraints proper to generative apparatuses require that no one claims "having an idea" of which he would strive to persuade others. The *interinflections* among partial and thus divergent perspectives are made in a tentacular mode, through reciprocal affections, without dramatic moments in which agreement would be imposed at the same time as reasons for it. There is no drama here because the reasons for agreement are the situation itself as it has received the power of making sense in common, thanks to the constraints associated with the apparatus.

Still, ideas cannot to be excluded. What is fearsome is the furious hold they have over those who serve as their mouthpieces. When beings or entities for whom no one can claim authority to speak are at stake in the common situation, there is nothing incongruous about them participating in making sense in common through the mode of inflection of utterances prompted by their presence. To speak in the presence of such beings or entities is not to express oneself (e.g., "I have the right to defend my idea"); it is to speak otherwise, constrained by that presence. For, even if that presence is silent, what is made present is attentive, and its attention obliges those who speak to slow down, to let themselves be touched, modified in a mode activating the tentacular character of the situation.

When those who have tasted the furious power of ideas participate in generative apparatuses, such apparatuses are effective in demanding a use of language that renders present those, human or not, who are concerned by the decision to be made, implicitly or explicitly. Critics would call such usage artificial, as they might say of rituals. In effect, such usages are "facts of art." It is possible to speak of artifice in the pejorative mode only if another usage of language is considered natural, one privileging the active voice, which expresses an idea furiously demanding realization.

At the outset, I noted how the manner in which Whitehead characterized humans in terms of foolish enterprises and dreams—they "would have crossed the Rubicon"—brought a frown to Haraway's face. In fact, when the Whiteheadian zigzag activates the middle voice, another figure now comes to mind instead of "the human": the *entrepreneur*. I see the entrepreneur as someone who defines the world in terms of how it poses problems, and even creates obstacles, for the realization of an idea. For the entrepreneur, anything that troubles the capacity for argumentation, deduction, or rational persuasion is deemed illegitimate or "artificial." The entrepreneur is animated by a possibility that requires him to disregard anything that might compromise the realization of that possibility. The entrepreneur is not guilty. Possibility is not to be prohibited. Ideas are not to be denounced. The active voice is not to be hunted down. What prove poisonous, however, are claims of innocence and legitimacy, which are vectors of anesthesia, and in fact intrinsically dangerous when it comes to learning to live in the ruins.

"Staying with the trouble," then, to use Haraway's motto, might be what a life in the ruins requires of what has been defined as civilization. Trouble is now everywhere, and however capable we are of knowing everything or nearly everything about what has devastated our worlds, that knowledge does not define the "solution," for devastation and reparation or regeneration are not symmetrical. To unmake is a rather easy undertaking. And unmaking may even remain blind to what it is destroyed, advancing in the name of a general good will or a captivating possibility. In contrast, to regenerate is never a general matter, for it is about creating or reactivating, step by step, relationships that are always tentacular, always partial, always to be cultivated, to be resumed under the aegis of the absence of guarantee, and also under the aegis of sorrow when loss is irreparable.[51]

Tentacular reality has nothing enchanting about it. Life may not be mere robbery; it may be the creation of symbioses, the composition of heterogeneous beings, the joy of interdependence; yet it is also capture and parasitism. But it is in that way, and not as purely and simply good, that life may afford motives for histories that we may prove capable of transmitting to those to whom we bequeath a precarious future, without guarantee, a future that will not, in any event, be a fairy tale in which all's well that ends well. A future that does pose the question of what is required for lives worth living, even in the ruins.[52]

Such histories are already beginning to proliferate. Almost everywhere today, an old term has reappeared: "the commons." The commons are not to be confused with common or public goods and not to be reduced to joint management of a resource that reckless and irresponsible use might destroy. To be sure, the fact that those who deal with such a resource are capable of self-governance, a mode of intelligent cooperation without needing externally imposed rules is in itself an achievement whose cultivation will be invaluable in the ruins. And it is indeed the middle voice that is at stake in the statement "there is no commons without commoning," no commons without making in common, without practices that entangle people, that make commoners of them.[53] But this making in common does not define commons as exclusively human. What sensitivity can making in common awaken in those who participate? With whom will they become capable of making in common? Who will participate? With whom will composition happen? The answers to such questions are surely not within the purview of philosophy. The answers can only be local, situated, practical, and if the question itself can be cultivated, relayed, passed on, it is by the wealth of circulating stories that activate the welding of common sense and imagination.

Whitehead can accompany us in the ruins. It is no doubt difficult for academic philosophers to admit that his is a "serious" philosophy, with arguments worth tackling in their dissertations. The motif I have attempted to bring forth with the strings proposed by Whiteheadian metaphysics will not reduce this difficulty—rather to the contrary. For the middle voice whose importance I have tried to stress is poorly suited, not only to historians of philosophy, but also to contemporary philosophy as "positioning" within an academic market where the philosopher-entrepreneur insists on the value of "his" concepts. "As we think, we live," wrote Whitehead,[54] and I am not sure how this old tradition called philosophical thought can contribute to life in the ruins. Of one thing I am certain: it will no longer be to philosophy that common sense will pose the question of coherence. It will fall to each "making sense in common" to reinvent the sense of this question. Perhaps, however, Lorde's warning that "the tools of the master will never dismantle the master's house" is what philosophy can bring to life in the ruins: philosophy that has tasted and tired of all the poisons fabricated by masters, that has explored all the variations on "it is this or otherwise chaos, arbitrariness, violence, treason." Undoubtedly, we will never be done with the tentacular capture by the idea that demands to be realized. Perhaps we still have need of philosophy to learn to taste, with a sense of humor that does not bear insult, the passions of what Whitehead called the "adventures of ideas."

Notes

Translator's Introduction

1. On Stengers and the refusal of mobilization, see also Bordeleau and van Tuinen, "Isabelle Stengers."
2. Stengers, *Invention of Modern Sciences*, 14.
3. Whitehead, *Aims of Education*, 107; cited in Stengers, *Invention of Modern Sciences*, 169n10.
4. Stengers, *Réactiver le sens commun*, 7–8. The author opted not to include her brief foreword in this translation.
5. Latour, "Foreword: Stengers's Shibboleth," xii.
6. Stengers, *Cosmopolitics II*, 363–71; see also Bordeleau and van Tuinen, "Isabelle Stengers."
7. Whitehead, *Process and Reality*, 27; see also 277.

1. The Question of Common Sense

1. Whitehead, *Modes of Thought*, 173–74.
2. Whitehead, *Modes of Thought*, 2.
3. Whitehead, *Modes of Thought*, 51.
4. Whitehead, *Modes of Thought*, 171–72.
5. Whitehead, *Modes of Thought*, 168–69.
6. Whitehead, *Modes of Thought*, 173.
7. Whitehead, *Process and Reality*, 17.
8. Rescher, *G. W. Leibniz's Monadology*, 233.
9. Bensaude-Vincent, *L'Opinion publique de la Science*.
10. Whitehead, *Process and Reality*, 74.
11. Whitehead, *Science and the Modern World*, 4.

12. Weinberg, *Dreams of a Final Theory*, 57–58
13. Brecht, "On Violence," in *Poems*, 276.
14. Whitehead, *Modes of Thought*, 29.
15. Baruk, *L'Âge du capitaine*.
16. Whitehead, *Science and the Modern World*, 59.
17. Whitehead, *Modes of Thought*, 49.
18. Whitehead, *Modes of Thought*, 50.
19. Whitehead, *Modes of Thought*, 116.
20. Whitehead, *Process and Reality*, 17.
21. Whitehead, *Science and the Modern World*, 196–97.
22. Whitehead, *Science and the Modern World*, 203.
23. I strived to follow this trajectory in Stengers, *Thinking with Whitehead*. The result is a rather dense book. For a lucid and elegant presentation of Whitehead's system, see Debaise, *Nature as Event*.
24. Whitehead, *Adventures of Ideas*, 273.
25. Whitehead, *Adventures of Ideas*, 274.
26. Whitehead, *Adventures of Ideas*, 164.
27. Whitehead, *Adventures of Ideas*, 164.
28. Whitehead, *Modes of Thought*, 27.
29. Whitehead, *Modes of Thought*, 30.
30. Whitehead, *Modes of Thought*, 26.
31. Whitehead, *Modes of Thought*, 111.

2. In the Grip of Bifurcation

1. Whitehead, *Science in the Modern World*, 56.
2. Whitehead, *Concept of Nature*, 20.
3. These extraterrestrials, passing close to the earth during their touristic wanderings, would perhaps see it in the mode proposed by a cartoon published in the press for the centenary of Einstein's birth: the planet Earth appears with other planets among the stars, but bearing a commemorative plaque on which is written: "Birthplace of Albert Einstein."
4. Whitehead, *Science in the Modern World*, 11.
5. Whitehead, *Modes of Thought*, 79–80.
6. Whitehead, *Modes of Thought*, 99.
7. Whitehead, *Modes of Thought*, 80.
8. Whitehead, *Modes of Thought*, 136.
9. Whitehead, *Modes of Thought*, 133.
10. Whitehead, *Process and Reality*, 171.
11. Whitehead, *Modes of Thought*, 121.

12. Whitehead, *Modes of Thought*, 135.
13. Whitehead, *Process and Reality*, 52, 223.
14. Whitehead, *Religion in the Making*, 96–97.
15. Rose, "Science Wars."
16. Gould, *The Mismeasure of Man*.
17. Whitehead, *Science in the Modern World*, 52.
18. Whitehead, *Modes of Thought*, 8.

19. It is worth noting that this importance goes so far as to justify that such arbitration be required from beings that make their milieu matter (animal and human experimentation) as if they were indifferent to the maddening senseless character of the milieu that experimenters impose on them. The myth of finite facts enables the production of torturers.

20. The image of justice, eyes blindfolded, a sword held in one hand and a balance in the other, translates a practice very different from formal, mathematical, and logical practices. Justice *must not* see, must not take into account that whose omission its facts demand, to ensure that the balance, generator of commensurability, continues to function, and that the sword, synonymous with conclusion, may slice. For an account of the ways in which the blindfold is negotiated, the definition of what will be included and omitted, and the operation of the balance, see Latour, *Making of Law*. Logic and mathematics have no need for a blindfold; they see only what their premises permit them to see.

21. See, for instance, Stengers, *La Vierge et le Neutrino*.
22. Weinberg, *Dreams of a Final Theory*, 28–29.
23. On this topic, see Stengers, *Invention of Modern Science*.

24. Several years were required for Alice to escape this sense of doom, and it is this experience that led to the creation of the collective Dingdingdong, at once both to sound the alarm and to give voice to other ways of knowing and telling, as well as experiences and propositions that allow one to envision the genuine possibility of living a life worth living with Huntington's disease. See the web source dingdingdong.com, as well as Sohldju, *Testing Knowledge* (which is preceded by Rivières's *Dingdingdong Manifesto*).

25. Schneider, *Science as a Contact Sport*.

26. New types of history are currently being written on this subject. See, for instance, Patel and Moore, *History of the World*.

27. This accounts for the perplexity when we encounter the manner in which Thomas Aquinas, for instance, creates an agreement among the authors he cites with respect to a point in dispute by shamelessly misrepresenting them. It is because these Authors are "authorities," and one authority cannot, in fact, contradict another. It is a matter of letting

everyone of them be right. Print, which dethrones authorities, reducing Authors to the status of authors endowed with opinions, would destroy this art, and would reinstate polemic as the "natural" law for debates.

28. See Starhawk, *Dreaming the Dark*.

29. Jacques Testart and the association known as Citizens' Sciences in France promoted the term "Citizen's Convention." The conventions respond to a precise protocol the respect of which is crucial for the apparatus not to sink into the flows of "participatory democracy," which is also to say, not to become an instrument aiming to ensure the public "acceptability" of an innovation, as is often the case today.

30. It is one of the most noteworthy constant features of Citizen's Conventions—for instance, of a recent one in France around the disposal of nuclear waste—that those who have been asked to think collectively about a proposition entailing risks resist the reassurances of experts who strive to circumscribe these risks and to define them as "manageable." They demand that these risks be taken seriously and, in the case of nuclear waste, insist for instance on the obligation to imagine the means for humans to keep the needed vigilance for the coming millennia.

31. Whitehead, *Modes of Thought*, 78.

32. Whitehead, *Modes of Thought*, 111.

3. A Coherence to Be Created

1. Whitehead, *Modes of Thought*, 83.
2. Whitehead, *Modes of Thought*, 62.
3. Whitehead, *Modes of Thought*, 78.
4. Whitehead, *Science in the Modern World*, 77–78.
5. Despret, *What Would Animals Say If We Asked the Right Questions?*
6. A recollection here comes to me from the time when I was learning to understand the significance of users' movements, which meant in that instance users who had neither quit nor were trying to quit illegal drugs. I especially recall my anger upon hearing a sociologist highlight the "constructed" character of what he could obtain from the actors in these movements, which, he said, was far less rich in nuance, contradictions, and ambiguities than the testimony of "mere consumers." I recall spitting in his face: "But who do you think you are to extract such confessions! That isn't sociology, it's voyeurism!"
7. Whitehead, *Modes of Thought*, 109.

8. Whitehead, *Modes of Thought*, 111.
9. Whitehead, *Science in the Modern World*, 157–58.
10. Leibniz, *Confessio philosophi*, 74–75. See also Leibniz's *New Essays on Human Understanding*, bk. 2, ch. 21, §47.
11. Whitehead, *Modes of Thought*, 105.
12. Debaise, *Nature as Event*.
13. Whitehead, *Process and Reality*, 86.
14. Whitehead, *Process and Reality*, 222.
15. For a devastating account of the impasses this biology confronts, see Kupiec and Sonigo, *Ni Dieu ni gène*. It should be noted that the critique of the bifurcation between end and cause is already developed by Henri Bergson in *Creative Evolution* and by Raymond Ruyer.
16. The French translation of *Process and Reality* translates "aim" both as *visée* and *but*, indifferently. It is worth noting that *but*, or "goal," introduces a kind of abstraction that should be avoided.
17. In fact, "prehension" finds a near equivalent in "feeling" in *Process and Reality*, with the caveat that feeling must be felt while, as we shall see, the price of achievement is that certain prehensions can be excluded from feeling. I will subsequently utilize the one term or the other in accordance with the context imposed by the citations.
18. In *Thinking with Whitehead*, I tried to follow the manner in which Whitehead understood these exigencies. See in particular ch. 18, "Feeling One's World." See also Debaise, *Nature as Event*.
19. Whitehead, *Process and Reality*, 85. Whitehead is reworking the prophecy of Ezekiel 37:10. Here I am taking the liberty of building on what I developed in *Thinking with Whitehead*, 298 and 385.
20. Whitehead, *Process and Reality*, 226–27.
21. Whitehead, *Adventures in Ideas*, 177. Lucretius, *On the Nature of Things*, 29: "Then again, just suppose that all the existing space were finite, and that someone ran forward to the edge of its farthest border and launched a spear into flight: do you favor the view that the spear, cast with virile vigor, would fly far and reach its target, or do you suppose that something could check it by obstructing its course?"
22. Deleuze and Guattari, *What Is Philosophy?*, 47.
23. Whitehead, *Modes of Thought*, 165–66.
24. I do not claim any expertise in Latin, for, in contrast with Whitehead, I have none. Here I draw heavily on Couffin, "Les verbes significant 'penser' chez Plaute et Terence."
25. Whitehead, *Modes of Thought*, 99.
26. Whitehead, *Process and Reality*, 35. It is impossible to insist too

much on the invaluable possibilities of the terms "character" and "characterize."
    27. Whitehead, *Modes of Thought*, 115.
    28. Whitehead, *Process and Reality*, 107.
    29. Whitehead, *Process and Reality*, 35.
    30. Whitehead, *Modes of Thought*, 97.
    31. Whitehead, *Modes of Thought*, 97.
    32. Whitehead, *Modes of Thought*, 98.
    33. See Stengers, "Invention of Mechanics," in *Cosmopolitics I*.
    34. For more on this subject, see the appendix "Le premier dispositive experimental?" in Stengers, *La Vierge et le Neutrino*.
    35. Whitehead, *Modes of Thought*, 154.
    36. Whitehead, *Modes of Thought*, 138.
    37. Whitehead, *Modes of Thought*, 147.
    38. And quantum mechanics are included here because the famous quantum wave-function evolving in Hilbert space also responds to the ideal of self-sufficiency. It is its "reduction," necessary to the definition of the probability of observing measurable magnitudes, that violates this ideal (cf. the so-called "measurement problem"). This infraction made room, and still makes room, for literally unbridled interpretations (direct intervention of the knowing human mind, worlds multiplying from reduction to reduction, etc.), as if all were permitted in order to avoid seeing physics become one science among others, relative like the others to the questions it has successfully obtained answers to.
    39. Whitehead, *Modes of Thought*, 150.
    40. Whitehead, *Modes of Thought*, 144.
    41. Whitehead once gave this advice to his students: "Meditate on your guts" (Weber, "The Organic Turn," 110).
    42. Whitehead, *Modes of Thought*, 150.
    43. Lorde, "The Master's Tools Will Never Dismantle the Master's House," in *Sister Outsider*.
    44. Whitehead, *Modes of Thought*, 161.
    45. Whitehead, *Modes of Thought*, 167.

4. What Can a Society Do?

    1. See Stengers, *Cosmopolitics II* and *La Vierge et le Neutrino*.
    2. Latour, *Inquiry into Modes of Existence*, 59.
    3. Latour, *Inquiry into Modes of Existence*, 59.
    4. Whitehead, *Process and Reality*, 79.
    5. See Stengers, *Another Science Is Possible*.

6. See ch. 1, "Toward a Public Intelligence of the Sciences," in Stengers, *Another Science is Possible*.

7. Williams James, "Will to Believe," in *Will to Believe*, 3. The French translation proposes *parfaite* for "genuine," which I find hard to accept. "Veritable" or "authentic" would have been far more adequate.

8. Whitehead, *Concept of Nature*, 40.

9. Whitehead, *Concept of Nature*, 29.

10. Whence the tension between psychotherapeutic arts and scientific reference. In *Hypnosis between Science and Magic*, I argued that the fundamental indifference Freud attributed to the unconscious is what permitted him to claim to decode its effects without those effects being open to counterinterpretation on the basis of suggestion.

11. Technical success is something else entirely. It may be placed under the sign of an irreducible coresponsibility. In *The Mangle of Practice*, Andrew Pickering compares the development of a detection apparatus to a dance. To be sure, the one who develops such an apparatus is supposed to be able later to give an account of what he has done by referring to what is detected as the only thing responsible for its detection. But it is only at the end of a two-stage dance. In the first stage, the human acts, regulates, develops; in a second stage, he lets the detector function and observes how it responds to the events affecting it. Then he returns to his active role of agent before letting the detector act again, until the detector valorizes in a stable and univocal manner what it is its function to valorize; it "reacts" only to those events for the detection of which it has been developed. The detector now has the value of an instrument thanks to which the experimenter may obtain reliable information about what is detected. Success, here, is that of "making do," not of a proof. See Stengers, "Life and Artifice," in *Cosmopolitics II*. For the entanglement of humans and techniques mobilized by the proof, see, too, the excellent book by Myers, *Rendering Life Molecular*.

12. Whitehead, *Modes of Thought*, 28–29.

13. Whitehead, *Modes of Thought*, 22.

14. See, on the power of the "=" sign, Stengers, *Cosmopolitics I* (116–19).

15. Whitehead, *Modes of Thought*, 78.

16. Whitehead, *Modes of Thought*, 51.

17. Chemical substances are considered as intelligible by quantum physics, but only in the sense in which the representation of molecules, stemming from delicate negotiations between quantum intelligibility and experimental data, allows for understanding their possibilities of composition and decomposition, not in the sense of submitting these

events as such to physics. Correlatively, chemists cannot avoid speaking of chemical substances as "actors," making of them syntactic subjects, not terms of a function.

18. See in Gilles Deleuze and Félix Guattari's *What Is Philosophy?* the related notion of the partial observer that, for its part, experiences effects but does not act (129–33).

19. Latour, *Facing Gaia*, 170.

20. Selectionist biology also indulges in overanimation, but it does so by turning the "aim to serve" into what would be its true cause, natural selection, which is supposed to be the unique source of all intelligibility. On this topic, see Pierre Sonigo's chapter "Cellules en liberté" in Kupiec and Sonigo, *Ni Dieu ni gène*. Sonigo proposes to think the body as a forest, which permits him to make a mockery of the idea that all individuals participating in a forest are intelligible through the manner in which they serve the forest, and to lend an attentive ear instead to what foresters know when they say "the forest is sick," accepting quite happily that trees depend vitally on each other and on many other living beings. The analogy of the body with a forest does not permit throwing out Darwinian selection, of course, but it requires not making it into a magic formula that resolves all problems.

21. It goes without saying that "competence" here has nothing to do with the impoverished theoretical abstraction of the same name, which is supposedly definable independently of the situation and verifiable in general in the mode of "being capable of" (see Stroobants, "Transduction").

22. On this contrast, see the remarkable book by Keller, *Making Sense of Life*.

23. Deleuze and Guattari, *Thousand Plateaus*, 52.

24. Kauffman, *Investigations*, 151.

25. "Exaption" is a term created by Gould and Vrba for designating—in contrast with adaptation—cases in which a trait takes on meaning or endorses a role that it had not been selected for; the pressure of selection comes only afterwards. The genealogy of what today is the "wing" for birds and the genealogy of mammals' milk are classic examples (see Gould and Vrba, "Exaption").

26. Stuart Kauffman, *Investigations*, 134–35.

27. Whitehead, *Process and Reality*, 105.

28. Rescher, *G. W. Leibniz's Monadology*, 233.

29. Whitehead, *Process and Reality*, 105.

30. Or: "How, *here, in this site*, can we learn to live together with

these others who disturb us but of whom we also know that, in their own mode, they are capable of paying attention to what we do?"

31. Gilbert, Sapp, and Tauber, "Symbiotic View of Life." This article was a manifesto, and in its wake, the number of books with a large amateur public has rapidly multiplied.
32. Gilbert et al., "Symbiosis as a Source of Selectable Epigenetic Variation," 673.
33. Haraway, *When Species Meet*, 227. See, too, Haraway, *The Manifesto of Companion Species*.
34. Haraway, *When Species Meet*, 215.
35. Haraway, *When Species Meet*, 224.
36. Whitehead, *Modes of Thought*, 76–77.
37. Whitehead, *Process and Reality*, 104.
38. Whitehead, *Process and Reality*, 105–6.
39. Whitehead, *Process and Reality*, 105.
40. Haraway, *When Species Meet*, 240: "Play is not making a living; it discloses living."
41. Haraway, *When Species Meet*, 219.
42. Hustak and Myers, "Involuntary Momentum."
43. Deleuze and Guattari, *Thousand Plateaus*, 238.

5. A Metamorphic Universe

1. Haraway, *Staying with the Trouble*, 33, 58–98.
2. Haraway, *Staying with the Trouble*, 173n2.
3. It suffices to consult websites for soap opera fans to be impressed by the passion with which is denounced any arbitrary decision by the scriptwriters who decide to change the behavior of a character without making the effort to make the transformation understandable, that is shareable (or "infectious").
4. Whitehead, *Process and Reality*, 11.
5. It has been said that Whitehead was the least read but most cited philosopher, and it is undoubtedly a testament to the efficacy of this exercise of his thought. Readers who encounter such citations and reprise them do so not in order to promote the authority of the philosopher, but because they feel "touched" and know others will be too.
6. The idea that "it may fail" and that impatience is a factor in such failure is central to the concept of instauration proposed by Étienne Souriau in *Different Modes of Existence*. See, too, Stengers, *Thinking with Whitehead*, 462–65. Souriau would undoubtedly recognize the

task that Whitehead gave to philosophy as an instauration, the response to the solicitation of facts to not be considered as simple facts, but to take on the power of being perceived as achievement, testifying to the solemnity of the world.

7. Schillmeier, *Eventful Bodies.*
8. Whitehead, *Process and Reality*, 187.
9. Whitehead, *Modes of Thought*, 36.
10. Whitehead, *Process and Reality*, 235.
11. Whitehead, *Process and Reality*, 222.
12. Whitehead, *Modes of Thought*, 70.
13. Whitehead, *Process and Reality*, 85. Whitehead cites the vision of Ezekiel 37:10 describing the miracle of resurrection, but for Whitehead, it's the creation of a new being that is at stake.
14. Bouteldja, *Whites, Jews, and Us.*
15. Whitehead, *Science and the Modern World*, 20.
16. See Benveniste, "Active and Middle Voice in the Verb," in *Problems in General Linguistics*, 145–52.
17. Derrida, *Margins of Philosophy*, 9.
18. Latour, "Factures/Fractures."
19. Whitehead, *Process and Reality*, 222.
20. Whitehead, *Modes of Thought*, 165.
21. Whitehead, *Adventures of Ideas*, 176.
22. Haraway, *When Species Meet*, 236.
23. Whitehead, *Process and Reality*, 226–27. Quakers understand this: the gatherings through which they make collective decisions require each and every one to recall that they "may be mistaken" but they make decisions all the same.
24. Stroobants, "Transduction."
25. Whitehead, *Modes of Thought*, 41.
26. Whitehead, *Modes of Thought*, 62
27. Whitehead, *Adventures of Ideas*, 148.
28. Whitehead, *Adventures of Ideas*, 148.
29. See Deleuze and Guattari, *What Is Philosophy?*, esp. 9–10, about rivalry in relation to this milieu of friends and equals.
30. Haraway, *Staying with the Trouble*, 30–57.
31. Abram, *Spell of the Sensuous*, 51.
32. Haraway, *Staying with the Trouble*, 34.
33. Viveiros de Castro, *Cannibal Metaphysics*, 40.
34. Winthereik and Verran, "Ethnographic Stories as Generalizations that Intervene."

35. James, *Psychology*, 387–88.

36. See Martin, *Les Âmes sauvages*, for a nonromanticized approach to what we call animism. Martin does not describe animist good savages endowed with a native intelligence of natural equilibria, but deciphers the problematic character of relationships that the hunter entertains with his prey, metamorphic relationships that can always switch into fascination, into invasion, into indistinction between self and other, into "alienation" (not in the critical sense, but rather in the sense in which, in the nineteenth century, those doctors whom we now call psychiatrists were alienists). The Gwich'in people in Alaska today are now caught between climate change and extractive industries, mining and oil, and the management projects of "defenders of nature," but their world has never been a stable "traditional" world; it is a world in which metamorphoses, uncertainty, and the threat of destruction are inherent.

37. Macknik, Martinez-Conde, and Blakeslee, *Sleights of Mind*.

38. Abram, *Spell of the Sensuous*, 41

39. Abram, *Spell of the Sensuous*, 83.

40. Whitehead, *Modes of Thought*, 37.

41. Whitehead, *Modes of Thought*, 38.

42. Deleuze, *Difference and Repetition*, 219.

43. It is to my long-time accomplice Olivier Hofman that I owe the idea of precariat thought as the art of living without the protection of what claims to think for us. "Precarious as we live on an Earth that is precarious itself, we can feel and be felt, taste and be tasted, see, hear, touch, and be seen, heard, touched. We pass through a Milieu that passes through us."

44. Haraway, *Staying with the Trouble*, 9–29

45. Whitehead, *Science and the Modern World*, 111–12 and 205–6.

46. On this topic and others, see Danowski and Viveiros de Castro, *Ends of the World*.

47. Danowski and Viveiros de Castro, *Ends of the World*.

48. Tsing, *Mushroom at the End of the World*. It must be noted that the ruins left by the operations of capitalist extraction are not, here, synonymous with the eradication of life: it happens that they are opportunity, making way for new zones of contact, for transformative encounters that let a "more than human" sociality emerge, polyphonic assemblages in which humans take part but not designers. But Tsing warns us that salvage capitalism, unlike state bureaucrats, has no fear of ruins.

49. See Stroobants, "Transduction."

50. Whitehead, *Modes of Thought*, 119.
51. See Van Dooren, *Flight Ways*.
52. William James addresses this question in his essay "Is Life Worth Living?" in *Will to Believe*.
53. See Bollier, *Think Like a Commoner*.
54. Whitehead, *Modes of Thought*, 63.

Bibliography

Abram, David. *The Spell of the Sensuous: Perception and Language in a More-Than-Human World.* New York: Vintage, 1997.
Baruk, Stella. *L'Âge du capitaine: de l'erreur en mathématique.* Paris: Seuil, 1985.
Bensaude-Vincent, Bernadette. *L'Opinion publique de la Science: à chacun son ignorance.* Reprint with unpublished afterword. La Découverte/Poche 392. Paris: Decouverte, 2013.
Benveniste, Émile. *Problems in General Linguistics.* Miami Linguistics 8. Coral Gables, Fla.: University of Miami Press, 1971.
Bollier, David. *Think Like a Commoner: A Short Introduction to the Life of the Commons.* Gabriola Island, BC, Canada: New Society, 2014.
Bordeleau, Erik, and Sjoerd van Tuinen. "Isbelle Stengers: An Introduction." In *De nieuwe Franse filosofie: denkers en thema's voor de 21e eeuw,* edited by Bram Ieven, Aukje van Rooden, Marc Schuilenburg, and Sjoerd van Tuinen, 440–52. Amsterdam: Boom, 2011.
Bouteldja, Houria. *Whites, Jews, and Us: Toward a Politics of Revolutionary Love.* Translated by Rachel Valinsky. Semiotext(e) Intervention 22. South Pasadena, Calif.: Semiotext(e), 2017.
Brecht, Bertolt. *Poems.* Edited by John Willett and Ralph Manheim. London: Eyre Methuen, 1976.
Couffin, Jeanne. "Les verbes significant 'penser' chez Plaute et Terence." In *Les problèmes de la synonymie en latin,* edited by Claude Moussy, 141–42. Lingua Latina. Paris : Sorbonne University Press, 1994.
Danowski, Déborah, and Eduardo Batalha Viveiros de Castro. *The Ends of the World.* Cambridge: Polity, 2017.
Debaise, Didier. *Nature As Event: The Lure of the Possible.* Translated

by Michael Halewood. Thought in the Act. Durham, N.C.: Duke University Press, 2017.

Deleuze, Gilles. *Difference and Repetition.* Translated by Paul Patton. New York: Columbia University Press, 1994.

Deleuze, Gilles, and Guattari Félix. *A Thousand Plateaus.* Vol. 2 of *Capitalism and Schizophrenia.* Translated by Brian Massumi. Minneapolis: University of Minnesota Press, 1987.

Deleuze, Gilles, and Guattari Félix. *What Is Philosophy?* Translated by Graham Burchell and Hugh Tomlinson. European Perspectives. New York: Columbia University Press, 1994.

Derrida, Jacques. *Margins of Philosophy.* Translated by Alan Bass. Chicago: University of Chicago Press, 1982.

Despret, Vinciane. *What Would Animals Say If We Asked the Right Questions?* Translated by Brett Buchanan. Minneapolis: University of Minnesota Press, 2016.

Gilbert, Scott F., Jan Sapp, and Alfred I. Tauber. "A Symbiotic View of Life: We Have Never Been Individuals." *The Quarterly Review of Biology* 87 (2012): 325–41.

Gilbert, Scott, Emily McDonald, Nicole Boyle, Nicholas Buttino, Lin Gyi, Mark Mai, Neelakantan Prakash, and James Robinson. "Symbiosis as a Source of Selectable Epigenetic Variation: Taking the Heat for the Big Guy." *Philosophical Transactions of the Royal Society B* 365 (2010): 671–78.

Gould, Stephen Jay. *The Mismeasure of Man.* New York: Norton, 1993.

Gould, Stephen Jay, and Elisabeth Vrba. "Exaption—A Missing Term in the Science of Form. *Paleobiology* 8, no. 2 (1982): 4–15.

Haraway, Donna J. *The Companion Species Manifest: Dogs, People, and Significant Otherness.* Paradigm 8. Chicago: Prickly Paradigm, 2003.

Haraway, Donna J. *Staying with the Trouble: Making Kin in the Chthulucene.* Experimental Futures. Durham: Duke University Press, 2016.

Haraway, Donna J. *When Species Meet.* Posthumanities. Minneapolis: University of Minnesota Press, 2007.

Hustak, Carla, and Natasha Myers. "Involutionary Momentum: Affective Ecologies and the Sciences of Plant/Insect Encounters." *differences* 23, no. 3 (2012): 74–118.

James, William. *Psychology: Briefer Course.* Edited by Frederick H. Burkhardt, Fredson Bowers, and Ignas K. Skrupskelis. The Works of William James (Electronic Edition) 14. Charlottesville, Va.: InteLex, 2008.

James, William. *"The Will to Believe" and Other Essays in Popular Philosophy and Human Immortality.* New York: Dover, 1956.

Kauffman, Stuart A. *Investigations.* New York: Oxford University Press, 2003.
Keller, Evelyn Fox. *Making Sense of Life: Explaining Biological Development with Models, Metaphors, and Machines.* Cambridge, Mass.: Harvard University Press, 2002.
Kupiec, Jean-Jacques, and Pierre Sonigo. *Ni Dieu ni gène.* Paris: Seuil, 2000
Latour, Bruno. "Foreword: Stengers's Shibboleth." In Stengers, *Power and Invention,* vii–xx.
Latour, Bruno, and Monique Girard Stark. "Factures/Fractures: From the Concept of Network to the Concept of Attachment." *RES* 36 (Autumn 1999): 20–31.
Latour, Bruno. *The Making of Law: An Ethnography of the Conseil d'etat.* Cambridge: Polity, 2010.
Latour, Bruno. *An Inquiry into Modes of Existence: An Anthropology of the Moderns.* Translated by Catherine Porter. Cambridge, Mass.: Harvard University Press, 2013.
Latour, Bruno. *Facing Gaia: Eight Lectures on the New Climatic Regime.* Translated by Catherine Porter. Cambridge: Polity Press, 2017.
Leibniz, Gottfried Wilhelm. *Confessio Philosophi: Papers Concerning the Problem of Evil, 1671–1678.* Edited by R. C. Sleigh, Brandon Look, and James H. Stam. The Yale Leibniz. New Haven, Conn.: Yale University Press, 2005.
Lorde, Audre. *Sister Outsider: Essays and Speeches.* Crossing Press Feminist Series. Berkeley, Calif.: Crossing, 2007.
Lucretius Carus, Titus. *On the Nature of Things.* Translated by Martin Ferguson Smith. Indianapolis, Ind.: Hackett, 2001.
Macknik, Stephen L., S. Martinez-Conde, and Sandra Blakeslee. *Sleights of Mind: What the Neuroscience of Magic Reveals about Our Everyday Deceptions.* New York: Henry Holt, 2010.
Martin, Natassja. *Les âmes sauvages.* Paris : La Découverte, 2015.
Myers, Natasha. *Rendering Life Molecular: Models, Modelers, and Excitable Matter.* Experimental Futures. Durham, N.C.: Duke University Press, 2015.
Patel, Raj, and Jason W Moore. *A History of the World in Seven Cheap Things: Guide to Capitalism, Nature, and the Future of the Planet.* Oakland: University of California Press, 2017.
Pickering, Andrew. *The Mangle of Practice: Time, Agency, and Science.* Chicago: University of Chicago Press, 2010.
Rescher, Nicholas. *G. W. Leibniz's Monadology: An Edition for Students.* Pittsburgh: University of Pittsburgh Press, 1991.

Rose, Hilary. "Science Wars, My Enemy's Enemy is—Only Perhaps—My Friend." *Social Text* 45–46 (1996): 61–80.

Schillmeier, Michael W. J. *Eventful Bodies: The Cosmopolitics of Illness.* Theory, Technology and Society. Farnham, UK: Ashgate, 2014.

Schneider, Stephen Henry. *Science As a Contact Sport: Inside the Battle to Save Earth's Climate.* Washington, D.C.: National Geographic, 2009.

Sohldju, Katrin, *Testing Knowledge: Toward an Ecology of Diagnosis, preceded by "The Dingdingdong Manifesto" by Alice Rivières.* Translated by Damien Bright. Goleta, Calif.: Punctum, 2021.

Souriau, Étienne. *The Different Modes of Existence.* Translated by Erik Beranek and Tim Howles. Minneapolis, MN: Univocal, 2015.

Starhawk. *Dreaming the Dark: Magic, Sex & Politics.* Boston: Beacon, 1988.

Stengers, Isabelle. *Another Science is Possible: A Manifesto for Slow Science.* Translated by Stephen Muecke. Cambridge: Polity, 2018.

Stengers, Isabelle. *Cosmopolitics I.* Translated by Robert Bononno. Posthumanities 9. Minneapolis: University of Minnesota Press, 2010.

Stengers, Isabelle. *Cosmopolitics II.* Translated by Robert Bononno. Posthumanities 10. Minneapolis: University of Minnesota Press, 2011.

Stengers, Isabelle. *Hypnosis between Science and Magic.* Translated by Andrew Goffey. New York: Bloomsbury USA Academic, 2018.

Stengers, Isabelle. *The Invention of Modern Science.* Translated by Daniel W. Smith. Theory Out of Bounds 19. Minneapolis: University of Minnesota Press, 2000.

Stengers, Isabelle. *Réactiver le sens commun: lecture de Whitehead eu temps de débâcle.* Les Empêcheurs de penser en rond. Paris: Seuil, 2000.

Stengers, Isabelle. *Power and Invention: Situating Science.* Translated by Paul Bains. Theory Out of Bounds 10. Minneapolis: University of Minnesota Press, 1997.

Stengers, Isabelle. *Thinking with Whitehead: A Free and Wild Creation of Concepts.* Translated by Michael Chase. Cambridge, Mass.: Harvard University Press, 2011.

Stengers, Isabelle. *La Vierge et le Neutrino: Les scientifiques dans la tourmente.* Les Empêcheurs de penser en rond. Paris: Seuil, 2006.

Stroobants, Marcelle. "Transduction: L'apprentissage comme métamorphose." In *Gestes speculatifs,* edited by Didier Debaise and Isabelle Stengers. Dijon: Presses du réel, 2015.

Tsing, Anna Lowenhaupt. *The Mushroom at the End of the World: On the Possibility of Life in Capitalist Ruins.* Princeton, N.J.: Princeton University Press, 2015.

Van Dooren, Thom. *Flight Ways: Life and Loss at the Edge of Extinction.* Critical Perspectives on Animals. Theory, Culture, Science, and Law. New York: Columbia University Press, 2014.

Viveiros de Castro, Eduardo. *Cannibal Metaphysics.* Translated by Peter Skafish. Minneapolis: Univocal/University of Minnesota Press, 2015.

Weber, Michael. "The Organic Turn." In *La Science et le Monde modern d'Alfred North Whitehead,* edited by François Beets, Michel Dupuis, and Michel Weber, 97–118. Frankfurt: Ontos, 2006.

Weinberg, Steven. *Dreams of a Final Theory.* New York: Pantheon, 1992.

Whitehead, Alfred North. *Adventures of Ideas.* New York: Free Press, 1961.

Whitehead, Alfred North. *The Aims of Education.* New York: Simon and Schuster, 1967.

Whitehead, Alfred North. *The Concept of Nature: Tarner Lectures Delivered in Trinity College, November 1919.* Cambridge: Cambridge University Press, 1930.

Whitehead, Alfred North. *Modes of Thought: Six Lectures Delivered in Wellesley College, Massachusetts, and Two Lectures in the University of Chicago.* Free Press paperback ed. Toronto: Collier-Macmillan Canada, 1968.

Whitehead, Alfred North. *Process and Reality: An Essay in Cosmology; Gifford Lectures Delivered in the University of Edinburgh during the Session 1927–28.* Corrected and edited by David Ray Griffin and Donald W Sherburne. New York: Free Press, 1978.

Whitehead, Alfred North. *Religion in the Making: Lowell Lectures, 1926.* New York: Fordham University Press, 1996.

Whitehead, Alfred North. *Science and the Modern World: Lowell Lectures, 1925.* New York: Free Press, 1967.

Winthereik, Brit Ross, and Helen Verran. "Ethnographic Stories as Generalizations That Intervene." *Science & Technology Studies* 25, no. 1 (2012): 37–51.

Isabelle Stengers is professor of philosophy at the Université Libre de Bruxelles. She is author of *Cosmopolitics I* (2010) and *Cosmopolitics II* (2011) and coauthor of *Women Who Make a Fuss* (2014), all published by the University of Minnesota Press.

Thomas Lamarre is professor of cinema and media studies and East Asian languages and civilizations at the University of Chicago. He is author of *Uncovering Heian Japan, Shadows on the Screen, The Anime Machine* (Minnesota, 2009), and *The Anime Ecology* (Minnesota, 2018).

(continued from page ii)

54 *Postcinematic Vision: The Coevolution of Moving-Image Media and the Spectator*
Roger F. Cook

53 *Bleak Joys: Aesthetics of Ecology and Impossibility*
Matthew Fuller and Olga Goriunova

52 *Variations on Media Thinking*
Siegfried Zielinski

51 *Aesthesis and Perceptronium: On the Entanglement of Sensation, Cognition, and Matter*
Alexander Wilson

50 *Anthropocene Poetics: Deep Time, Sacrifice Zones, and Extinction*
David Farrier

49 *Metaphysical Experiments: Physics and the Invention of the Universe*
Bjørn Ekeberg

48 *Dialogues on the Human Ape*
Laurent Dubreuil and Sue Savage-Rumbaugh

47 *Elements of a Philosophy of Technology: On the Evolutionary History of Culture*
Ernst Kapp

46 *Biology in the Grid: Graphic Design and the Envisioning of Life*
Phillip Thurtle

45 *Neurotechnology and the End of Finitude*
Michael Haworth

44 *Life: A Modern Invention*
Davide Tarizzo

43 *Bioaesthetics: Making Sense of Life in Science and the Arts*
Carsten Strathausen

42 *Creaturely Love: How Desire Makes Us More and Less Than Human*
Dominic Pettman

41 *Matters of Care: Speculative Ethics in More Than Human Worlds*
Maria Puig de la Bellacasa

40 *Of Sheep, Oranges, and Yeast: A Multispecies Impression*
Julian Yates

39 *Fuel: A Speculative Dictionary*
Karen Pinkus

38 *What Would Animals Say If We Asked the Right Questions?*
Vinciane Despret

37 *Manifestly Haraway*
Donna J. Haraway

36 *Neofinalism*
Raymond Ruyer

35 *Inanimation: Theories of Inorganic Life*
David Wills

34 *All Thoughts Are Equal: Laruelle and Nonhuman Philosophy*
John Ó Maoilearca

33 *Necromedia*
Marcel O'Gorman

32 *The Intellective Space: Thinking beyond Cognition*
Laurent Dubreuil

31 *Laruelle: Against the Digital*
Alexander R. Galloway

30 *The Universe of Things: On Speculative Realism*
Steven Shaviro

29 *Neocybernetics and Narrative*
Bruce Clarke

28 *Cinders*
Jacques Derrida

27 *Hyperobjects: Philosophy and Ecology after the End of the World*
Timothy Morton

26 *Humanesis: Sound and Technological Posthumanism*
David Cecchetto

25 *Artist Animal*
Steve Baker

24 *Without Offending Humans: A Critique of Animal Rights*
Élisabeth de Fontenay

23 *Vampyroteuthis Infernalis: A Treatise, with a Report by the Institut Scientifique de Recherche Paranaturaliste*
Vilém Flusser and Louis Bec

22 *Body Drift: Butler, Hayles, Haraway*
Arthur Kroker

21 *HumAnimal: Race, Law, Language*
Kalpana Rahita Seshadri

20 *Alien Phenomenology, or What It's Like to Be a Thing*
Ian Bogost

19 *CIFERAE: A Bestiary in Five Fingers*
Tom Tyler

18 *Improper Life: Technology and Biopolitics from Heidegger to Agamben*
Timothy C. Campbell

17 *Surface Encounters: Thinking with Animals and Art*
Ron Broglio

16 *Against Ecological Sovereignty: Ethics, Biopolitics, and Saving the Natural World*
Mick Smith

15 *Animal Stories: Narrating across Species Lines*
Susan McHugh

14 *Human Error: Species-Being and Media Machines*
Dominic Pettman

13 *Junkware*
    Thierry Bardini

12 *A Foray into the Worlds of Animals and Humans,*
    with *A Theory of Meaning*
    Jakob von Uexküll

11 *Insect Media: An Archaeology of Animals
    and Technology*
    Jussi Parikka

10 *Cosmopolitics II*
    Isabelle Stengers

9   *Cosmopolitics I*
    Isabelle Stengers

8   *What Is Posthumanism?*
    Cary Wolfe

7   *Political Affect: Connecting the Social and
    the Somatic*
    John Protevi

6   *Animal Capital: Rendering Life in Biopolitical Times*
    Nicole Shukin

5   *Dorsality: Thinking Back through Technology
    and Politics*
    David Wills

4   *Bíos: Biopolitics and Philosophy*
    Roberto Esposito

3   *When Species Meet*
    Donna J. Haraway

2   *The Poetics of DNA*
    Judith Roof

1   *The Parasite*
    Michel Serres